Research Projects
in
High School Biology

Research Projects
in
High School Biology

Dorothea Allen

Parker Publishing Company, Inc.
West Nyack, New York

PRINTED IN THE UNITED STATES OF AMERICA
ISBN–0-13-774422-6

To Bill

Creating an Atmosphere
of Biology-in-Action.....

Developing among high school students an absorbing and continued interest in the biological sciences requires an effective means of instruction. Moreover, it is important that the program provide dynamic experiences which will enhance the development of an understanding and appreciation of the nature and meaning of basic biological concepts.

This book offers the long-term research project as one method of creating an atmosphere of biology-in-action—an atmosphere in which students develop the necessary techniques for and scientific approach to the study of a problem through active involvement in its solution. The method makes good use of student interests and is readily adaptable to modern or traditional, beginning or advanced courses in high school biology. Using living materials and the element of experimentation in their study, students come to an understanding of how the sensitive and delicately balanced organization which we call life operates within equally complex surroundings.

Unlike "take-home" or "science-fair" projects, long-term research projects are conducted in the high school laboratory and make use of school time, materials and equipment. Unlike short-term laboratory investigations, they lend continuity to the laboratory studies and, being open-ended by nature, cast students in the role of biologists who work individually or in small groups to plan and conduct a biological study. During and after the study, discussion periods allow for comparison of results, identification of still unanswered questions, suggestions for further study, and, where indicated, formulation of basic plans for extended work.

Research Projects in High School Biology is a source book for teachers. It describes long-term projects and suggests how they can be used with high school biology classes. It presents an approach that does not pretend to be perfectly or completely planned, but rather leaves occasional procedures to the inventiveness of the particular teacher or student in his particular situation. It allows

9

for alternate procedures and encourages the use of wise judgment to turn some unexpected event or situation to good advantage.

Each chapter includes several project studies which center about a unifying theme. This grouping, however, places no restrictions on the teacher or students. On the contrary, recurring techniques to be employed and specimens recommended for study may suggest other effective groupings. Teachers are encouraged to select judiciously in accordance with expressed student interests, available materials, and depth of study deemed appropriate, while working toward the development of an individualized program of study tailored to suit each project situation.

The preliminary remarks which introduce each chapter give some historical background and evaluation of the importance of a particular field of investigation. They are included to help give some insight into the significance of a particular area of research and, together with the *motivation* introducing each project, to provide stimulation for student activity and involvement in the study of a real and challenging problem. Here, not only is the problem for study pointed out, but also the trend of exploration for its solution is, in general terms, defined.

Materials required are listed for each project so that the teacher can plan for the necessary quantities of materials and availability of same when needed. However, depending on the resourcefulness of the teacher, many improvisations and substitutions can be made with equal effectiveness, and often, depending on the locality, with greater convenience.

The *development* of individual research projects occupies a major portion of each chapter. The general approach outlines suggested procedures for the projects which, while delineated, allow for variations. Student ingenuity should be encouraged and many decisions concerning details are left to the discretion of the teacher.

Guidelines for *discussion and interpretation* follow each project development. These highlight the study and seek to develop broad implications of the field of inquiry, thereby giving a more fundamental meaning to the finished work. They may lead to unanswered questions, open-ended studies, or original work to be extended by individuals or by groups. They may suggest topics for written and/or oral reports relating to project findings and their importance to a problem of biological significance.

As a natural outgrowth of the discussion and interpretation of the project work, topics for *further study* may develop. Student interests should be a primary consideration in extending the project, but often variations of the basic outline can serve as a basis for comparison and give greater depth to the over-all project study. This section suggests some alternate procedures for each basic study as well as some open-ended projects suitable for individuals or small groups on a more selective basis.

As a final chapter section, *recommended reading* lists are included for a consultation of pertinent literature. It is suggested that students be made aware that, while time-consuming, library work is a needed prerequisite for research. Only through reading will they become acquainted with an understanding of a particular problem and its place in a more general field of biological endeavor. However, motivation for student reading should grow out of the laboratory work, and the references listed are to readable materials, many of which are suitable for students as well as their teachers.

The worth of the project activities may be evaluated by the questions students answer and by those which they raise; by their identification of unsolved problems and their development of scientifically sound plans for studying them; by their ability to express themselves clearly and effectively; and by the evidence they give which indicates the development of an active mind.

It is with pleasure that I gratefully acknowledge the helpful suggestions which have been gleaned from the many educators, biologists and curriculum study groups whose research has provided the basis for many of the ideas and techniques incorporated in *Research Projects in High School Biology*. My many high school classes also have been of invaluable assistance for they have worked with me to test, adapt, and re-test techniques, old and new, to the point of satisfactory performance on the high school level. It is with their help that I have discovered long-term projects to be practical, since they offer both exciting and rewarding experiences to high school biology students.

To achieve the ultimate—the laboratory as the central mechanism for learning biology—an ever-increasing integration of long-term projects may be the answer. To this end, it is hoped that teachers seeking ideas for suitable projects for their high school biology classes will find this book a source of interesting and helpful information.

GUIDELINES FOR THE TEACHER ....

The long-term project method is an important contribution to today's concept of high school biology instruction. It may be viewed as a developmental process in which there is a valuable impact for learning that is provided by the teacher—the teacher serves as a guide who stimulates and encourages students to active involvement in experimental study and helps them to develop the necessary skills for it; as experience with the technique grows, the student gradually works toward assuming greater responsibility for his own learning experiences; and, having been so guided and equipped in the preliminary phases of his development of an absorbing and continued interest in biology, the student embarks on what may rightly be classified as research-type projects.

Teachers, upon whom the successful employment of Research Projects in High School Biology rests so heavily, may find some assistance in the following guidelines:

• Initially, the entire content of a given topic chapter should be scanned so that a tentative selection of an appropriate project or combination of projects can be made for use, as stated or in some adaptive form. The project selected should be evaluated in terms of the biological process which it emphasizes, the laboratory techniques included which are already familiar to the students and those which must be developed, and its general suitability for the specific situation.

• The project should be developed with the students who will participate in its study. Student ideas should be encouraged and, when feasible, put into use. Often a variation or modification of the original project will prove to be more effective than the original, and students will identify more closely with a study which they have helped to develop than with one that is ready-made.

• Advance preparations should be made for each project study. *Materials required* for a given project indicates what will be needed for engaging in the basic project but substitutions and other modifications must also be considered. Suitable living specimens should be obtained and provisions for their culture and laboratory care made in advance of their arrival and the beginning of the project activity. Timing of the entire operation is of utmost importance, and week-end and vacation periods must be planned for, lest they interfere with project progress.

- It is desirable to illustrate, without supplying answers, the biological principles involved in a project. Some demonstrations associated with projects are included in the corresponding chapters, and teachers can draw on their own resources for many more that will help to stimulate and guide student thinking.

- Research reading should be encouraged to the fullest. By working cooperatively with the school librarian, both a classroom and a more extensive central school library of carefully selected scientific materials can be maintained for student reference reading. Applied to project findings and reported to the class, reference work will take on a new dimension, and students will find it a stimulating adjunct to the laboratory work.

- Communications concerning project findings should be permitted and planned for. Project results should be reported after the experimental work has been completed, and comparisons and interpretations should be made in concert. Data collected by students should be viewed by all with respect and, as an outgrowth of group evaluation of the results when compared with other pertinent information, re-testing or the formation of a new hypothesis and design of a new project to test it should be encouraged. For permanent reference, written records which follow an accepted format for research reports should be kept in student laboratory notebooks.

- The overlapping nature of techniques and methods used for various project studies should be noted and plans made for a continuing study of biological investigations that cross over the artificial boundaries set by chapter divisions. Capitalizing on the interest generated by a technique developed in one investigation by applying it as a device for another helps to broaden the student's perspective of the scope of biological investigation.

- Differential project assignments that recognize the enthusiasm of the student and direct his interests should be used to provide a highly individualized study program. The *further study* sections of each chapter offer suggestions for challenging the brighter and more highly motivated student and allowing him to engage in open-ended research projects of his own design.

- The program should be characterized by well-ordered flexibility which allows several related projects to be in progress at one time and for activity to be engaged in as suits the development of events. A rigid schedule which designates specific days for laboratory work and others for discussion periods should be avoided.

• An attempt to create an atmosphere of research and discovery in the laboratory should be made. A continuing teacher-research project should be in progress to set a stimulating example of biological investigation and allow students to observe techniques used, records kept, and progress made. Students should share in some of the routine operations associated with the teacher project and gradually be guided into projects of their own where they may experience the excitement of discovery through research in biology.

Dorothea Allen

SOURCEBOOKS FOR TEACHERS

Morholt, E., P. Brandwein, and A. Joseph. *A Sourcebook for the Biological Sciences,* Harcourt, Brace & World, New York, 1966.

Needham, James G. *Culture Methods for Invertebrate Animals.* Dover Publications, Inc., New York, 1937.

Pelczar, M. J., Jr. *Manual of Microbiological Methods of the Society of American Bacteriologists.* McGraw-Hill Book Book Company, New York, 1957.

Schwab, Joseph J. (ed.) *Biology Teachers' Handbook* (BSCS). John Wiley & Sons, Inc., New York, 1963.

Van Norman, Richard W. *Experimental Biology.* Prentice-Hall, Inc., Englewood Cliffs, New Jersey, 1963.

Table of Contents

1

ANTIBIOSIS

At all levels there is competition among living things. Organisms must compete for the available air, food, and space, and often some compete offensively for these vital factors by producing substances or conditions inimical to the normal growth or survival of other organisms. This phenomenon is known as *antibiosis.*

Antagonistic organisms, however, have no control over their production of antibiotics. The antibiotic substances produced are metabolic products which are formed regardless of any need of the organism, but since these products ward off competitors with no harmful effects to the producer, they provide the antagonist with an adaptive advantage. This is a basic concept to be developed early in the biology course.

Project studies involving antibiosis and the conditions which may enhance the activity of antibiotic substances are of value in establishing the nature of a competitive association and the specific advantage to which man can use his knowledge about it.

Project—NATURAL ANTAGONISMS BETWEEN MICROORGANISMS

MOTIVATION

Many microorganisms enter into an antagonistic association

known as antibiosis. Subtle in their tactics, some of the Actinomycetes produce substances known to interfere with the ribo-
somal activities of bacterial competitors, thus causing in them a
lack of essential proteins and, eventually, death. Others, equally
subtle, interfere with the synthesis of polypeptide cell walls of
competing bacterial populations. *Penicillium notatum,* a rather
well-known mold, is of the latter variety, using the action of its
secretions to render competitors defenseless and make them easy
prey to other natural enemies. Fortunately, the membranes of
man's cells are not constructed of the same molecules as are those
of the bacteria, hence penicillin can be used to counteract
bacterial infections in man while leaving the human cells and the
formation of their phospholipid membranes unaffected.

The substances derived from the offending members of antagonistic associations have potential value as antibiotic agents because
of their inhibitory effects against sensitive competitors. It is with
these inhibitors and their mode of action as they affect pathogens
in an antibiotic relationship that researchers are primarily concerned.

MATERIALS REQUIRED

A project which investigates the natural antagonisms between
microorganisms can be conducted with simple and relatively easily
obtained materials:

　　antagonist cultures—one 18 hour broth culture of each
　　　　Bacillus subtilis
　　　　Bacillus brevis
　　　　Pseudomonas aeruginosa
　　　　　　and one 10 ml. sterile water suspension of spores
　　　　　　from a 24 hour Sabouraud's dextrose agar slant of
　　　　Penicillium notatum;
　　test cultures—one 18 hour broth culture of each
　　　　Escherichia coli
　　　　Staphlococcus aureus
　　　　Sarcina lutea;
　　four sterile petri dishes;
　　four tubes, 15 ml. each, of sterile trypticase soy agar at 45° C;
　　an inoculating loop and flame;
　　masking tape; and

a transparent mm. ruler will be needed by each individual or group.

An incubator and an autoclave or pressure cooker should also be available.

DEVELOPMENT

Pitting one microorganism against another on the same growth surface may result in the inhibition of growth of one within the area into which secretions of the other have diffused. This identifies the former as the sensitive organism and the latter as the antagonist in an association of antibiosis.

Using sterile technique, each of four sterile trypticase soy agar plates should be poured, allowed to cool and solidify, and marked to identify the antagonist organism it is to receive. The appropriately marked plate should be inoculated by streaking one loopful of *Bacillus subtilis* broth culture in a straight line across the agar surface about one inch from the edge of the petri dish. In a similar manner, and taking precautions to guard against contamination, each of the remaining plates should be streaked with its designated antagonist culture. All plates should be taped with two inch strips of masking tape, inverted, and incubated at 37° C for 24-48 hours.

After incubation, the straight line growth of the organism on each plate should be examined and identified. Each plate should then be inoculated with the same three test organisms by streaking one loopful of each of the broth test cultures on the surface of each plate in a straight line and at right angles to the growth of the antagonist. Care should be taken to prevent contamination, and the same order of test organisms—left, *Escherichia coli;* middle, *Staphylococcus aureus;* right, *Sarcina lutea*—should be used for all plates which, when completed, will have an antagonist growth across the top and the three test organisms, well-spaced, parallel to each other and at right angles to the antagonist. Plates should be re-taped and again incubated at 37° C for 24-48 hours.

At the end of the incubation period, all plates should be examined for inhibition of growth of the test organisms. Where present, these distances should be measured with a mm. ruler and recorded on a suitably prepared chart. Interpretation of the results should be in the form of value judgments of "no antibiotic action," "moderate antibiotic action," or "strong antibiotic action" of each antagonist against each of the test organisms.

DISCUSSION AND INTERPRETATION

The results obtained by the various teams should be compared and evaluated. The terms "antibiosis" and "antibiotic" should be discussed and the relationship between them established. The value of antibiotic substances to the microorganisms producing them should be determined, as well as the value of these same substances to man.

The significance of the observation that some microorganisms produce antibiotics that are not effective against all other microorganisms should be noted, and from this a method for classifying antibiotic substances should be determined.

Reasons for the use of sterile technique—autoclaving of media, use of sterile equipment, flaming the inoculating loop, making all transfers aseptically—and the serious consequences that might ensue if these precautions were not taken should be stressed.

In each case where antibiotic activity was noted against sensitive test organisms, the antagonist should be researched to determine its "mode of action."

FURTHER STUDY

Variations of the basic project can be used to test natural antagonisms between other combinations of microorganisms. If desired, a more limited study can be made by using only two organisms at one time. From an 18 hour agar slant culture of *Sarcina subflava* a heavy suspension of the organism should be prepared in 45° C sterile trypticase soy agar, poured into a sterile petri dish and allowed to cool and solidify; the seeded surface should be inoculated in three well-separated areas with *Bacillus subtilis* from an 18 hour agar slant culture; plates should be taped, inverted, and incubated at 37° C for 24 hours; and an examination of the plates should be made to detect areas of inhibition of the *Sarcina subflava* growth around the three spotted colonies of *Bacillus subtilis* on the surface. Other organisms can be used, two at a time, in a similar manner.

Students should be encouraged to plan other projects which investigate the effects of antibiotic-producing microorganisms on the growth of higher forms of life. Some should be asked to design and conduct a project which tests the effect of a microorganism, shown to exert antibacterial action, on the growth of algae or young seedlings such as oat, radish, or lettuce. Others might plan an open-ended study to determine the effect of antibiotics on animal growth, using small animals in the laboratory for conducting their project. The basic project suggests many such extensions and allows for the use of both teacher and student ingenuity.

Project—WEED EXTRACTS VS. GERMINATION OF SEEDS

MOTIVATION

The success of a plant is by no means assured by its possession of suitable adaptations to the various factors of its physical environment. The presence of other plant species introduces the additional factor of competition and, lacking the physical attributes which an animal finds helpful while competing with others, the plant must use ingenious and devious methods to combat its competitors. Some exert an influence on the environment which, in turn, may interfere with the well-being of neighboring plants with which they are in competition; the production and excretion of metabolic substances by one plant may inhibit or retard the growth of others in the same area. The resulting relationship, antibiosis, may be conveniently studied by germinating seedlings of cultivated plants in petri dishes in which an extract of an antibiotic-producing plant is used in place of water.

MATERIALS REQUIRED

The potential of common weeds as producers of seed germination inhibitors may be studied conveniently with:

> 200 radish seeds;
> 12 whole fresh garden weeds;
> sterile water blanks, one each of 4 ml., 9 ml., 19 ml., 49 ml., 79 ml., 99 ml.;
> sterile pipettes, 1 ml. and 5 ml.;

eight sterile petri dishes lined with sterile filter paper circles,
   to fit;
2% solution of sodium hypochlorite;
1% detergent solution; and
a transparent mm. ruler for each run of the project.

The use of an autoclave and a Waring blendor, meat grinder or
mortar and pestle is also required.

DEVELOPMENT

Ten or twelve whole fresh common garden weeds should be
selected, washed, and macerated in a Waring blendor. The grinding
should be thorough and the mashed material filtered first through
double layers of cheesecloth to remove the pulp and then through
fine filter paper to remove all small particles. Solids should be
discarded and the filtrate autoclaved at 121° C for 15 minutes
under 15 pounds pressure. After sterilization, dilutions of the
filtrate should be made to yield extract concentrations of 1:5,
1:10, 1:20, 1:50, 1:80 and 1:100 in sterile distilled water.

With a sterile pipette, 5 ml. of the undiluted extract should be
transferred to a sterile petri dish lined with filter paper to fit.
Similarly, 5 ml. of each of the extract dilutions should be
transferred to its correspondingly marked paper-lined sterile dish.
An eighth dish, receiving 5 ml. of sterile distilled water, should be
prepared as a control.

Radish seeds that have been soaked successively for two
minutes in a 1% detergent solution, one minute in a 2% solution of
sodium hypochlorite, and one hour in sterile distilled water should
then be plated, 50 seeds to each prepared petri dish. Care should
be taken to avoid contamination; sterile forceps should be used for
transferring and placing the seeds on the surface of each plate and
the covers should be replaced promptly after the planting has been
completed.

After all germination chambers have been prepared they should
be placed in a cabinet or other area of reduced light at a
temperature of approximately 27° C.

Examination of the germination chambers should be made at 24
hour intervals. The number of seeds germinated and a calculation
of the percent of germination for each 24 hour period for each
weed extract concentration should be recorded on a suitably

prepared chart. These results should be compared with those of the control and the effectiveness of the various concentration levels of the weed extract as a seed germination inhibitor determined.

At the end of 72 hours, root and shoot growth of the seedlings should be measured in each of the concentrations of weed extract where germination was not inhibited. A comparison of the effects of the weed extract at various concentrations should be made, using the controls, grown in water, as a standard.

The project can be repeated with many variations. Various concentrations of the juice of ripened tomato fruit can be tested as a possible inhibitor of the germination of lima bean seeds, sweet clover hay extract can be tested against corn seed germination, and a study of the sap of burdock (*Arctium minus*) versus lettuce or radish seeds can be made.

DISCUSSION AND INTERPRETATION

To generate interest in the project and to promote discussion about it, the relative sensitivity of different seeds to the same antibiotic can be demonstrated. The germination of seven flax seeds, surrounded circularly by 20 seeds of the common falseflax, *Camelina sativa*, in a petri dish germination chamber can be compared with the germination of seven tomato seeds, surrounded similarly in a separate chamber. Controls, using no seeds of *Camelina sativa*, should be run in parallel.

If several variations of the basic project are used or if different seeds are assigned to each group using the same weed extract, discussion of the results should prove to be interesting. Results obtained may indicate that the same extract exerts complete inhibition over the germination of one type of seed and, at the same concentration, is totally ineffective against another.

The significance of plant antibiosis should be determined and, through careful analysis and interpretation of the data collected, suggestions, both general and specific, should be made for practical applications of this natural phenomenon. Consideration should be given to the project in explaining why tomato plants are never found growing in the vicinity of black walnut trees and why some other plants are not inhibited from growing in the same area.

FURTHER STUDY

Students who identify a weed extract as an effective seed germination inhibitor should be encouraged to test it against other seeds and, working within a narrower range, to determine its minimal inhibitory concentration for each type of seed tested. Those who may be curious about which of the weed plant organs are responsible for producing the inhibitory substance should be allowed to test extracts from the specialized plant organs—roots, stems and leaves—and to compare the results obtained with those observed when whole plant extract was used. Another extension of the project involves the use of extracts from plants representing different genera, pooled and tested to determine their relative effectiveness when used in combination and when used alone against the germination of a given type of seed. Other variations, such as the potential value of extracts from algae or lichens as inhibitors of seed germination, also offer challenging studies.

The more talented members within a group should be stimulated to plan and conduct a more advanced project. One which investigates the seed germination inhibitory activity of sap fractionates of common weeds or other plants can be run in parallel with the basic project or can serve as an open-ended study for selected students.

Project—ANTIBIOTICS VS. BACTERIA

MOTIVATION

The drug industry has been revolutionized in the last 25 years by the introduction of various antibiotics, yet bacteria differ greatly in their sensitivity and/or resistance to a particular antibiotic preparation. Mutations, which have provided the antibiotic-producing organism with an effective offense, may also be responsible for providing the target of the attack with an equally effective defense. Many cases of serious outbreaks of infections and diseases caused by resistant strains of bacteria have been difficult to control, primarily because the overuse of an antibiotic has eliminated the sensitive strains and left the resistant mutants free of competitors and unaffected by the compound which has remoyed their competition. A large initial dose of an effective antibiotic, when really needed, generally acts to control the greatest number of pathogens without harm to the patient and, at the same time, serves to prevent the build-up of resistant populations.

Studies of problems involving sensitivity and resistance of microorganisms to antibiotics contribute to a better understanding of bacterial populations and the changes they are likely to undergo. Further, they help to define the range of antibacterial activity of the many antibiotics now available to us.

MATERIALS REQUIRED

To determine the effects of various commercially prepared antibiotics on microorganisms, each group will need:

one 18 hour agar slant culture of each test organism:
Escherichia coli
Bacillus cereus
Staphylococcus aureus
Saccharomyces cerevisiae;
12 Difco Bacto-unidisks containing known amounts of eight of the more commonly employed antibiotics:
four unidisks with antibiotics at low concentration,
four with the same antibiotics at medium concentration,
and four of the same but at high concentration;
16 sterile petri dishes; and
four flasks of sterile trypticase soy agar, 50 ml. each.

An autoclave and an incubator should also be available.

DEVELOPMENT

A flask of melted trypticase soy agar, cooled to 45° C, should be inoculated heavily with *Escherichia coli* culture and shaken so as to form a uniform suspension. Allowing approximately 12 ml. per plate, the suspension should be poured into four sterile petri dishes and allowed to cool and solidify. Three Bacto-unidisks, each at a different concentration level, should then be transferred to the surface of individual poured plates, the fourth in the series serving as a control. The transfer of a unidisk from its sterile envelope should be made aseptically, using flamed forceps for the transfer and for tapping the disk gently but firmly to bring it in contact with the surface of the solidified agar suspension. The series of three plates should then be properly labeled to indicate the organisms in suspension and the relative concentration—high, medium or low—of the antibiotic on the unidisk. The control should be marked to indicate the test organism.

Repeating the same procedure, each of the other test organisms should be tested individually against the same concentrations of the same antibiotics. All plates should be incubated, upside down, at 37° C for 24 hours.

Following removal from the incubator, all plates should be examined for clear zones around the antibiotic-impregnated disks. These zones, representing areas of inhibition of growth of the test organisms, should be measured with a mm. ruler and the results tabulated on a chart designed to accommodate the kind and concentration of the antibiotic substances and their effectiveness as inhibitors of growth of the organisms tested.

DISCUSSION AND INTERPRETATION

Interpretation of the collected data should point up the importance of a zone of inhibition of growth as an indicator of the sensitivity of the test organism to the antibiotic. Different teams, assigned to test the antimicrobial activity of the same antibiotics at different concentration levels, should compare results and draw conclusions concerning the lowest inhibitory concentrations for specific organisms tested.

An obvious discussion point concerns the use of antibiotics as therapeutic agents. The concentration levels found to be effective against particular organisms should be evaluated and the reported side-effects attributed to the use of antibiotics should be investigated and discussed by the group. Students, no doubt, will have personal experiences as well to relate.

Whether it is desirable for an antibiotic to be bacterio-static or bactericidal should also be considered and the methods for determining which of these forms of action it exerts should be investigated. Application of information gathered from studies conducted *in vitro* to actual use *in vivo* has practical value also, and this should be discussed in relation to the project and its significance as a research study.

FURTHER STUDY

Students who identify a few pinpoint colonies growing in a clear zone should be encouraged to formulate an hypothesis about how they have developed. To test the hypothesis (e.g., the antibiotic has lost its potency, the colony has developed from a mutant, etc.) an individual or group should design and carry out an extension of the basic study.

Another project which defines the range of antimicrobial activity of specific antibiotic substances would be effective as a follow-up study for students who wish to pursue the topic further or as an alternate study for one or more of the better groups. Antibiotics with a broad antibiotic spectrum, such as aureomycin or terramycin (which are active at low concentrations against a large number of microbial species belonging to different genera), and those with a narrow antibiotic spectrum, such as penicillin (effective principally against only certain species of gram positive organisms), or polymyxin (effective against only certain species of gram negative organisms) should be used against a wide range of test organisms. Other antibiotics should also be used to determine the antibiotic spectrum for each and to suggest methods of classifying the antibiotic substances.

Project—ENHANCEMENT OF ANTIMICROBIAL ACTIVITY

MOTIVATION

Oligodynamic action is the seemingly lethal effect that certain metals in minute quantities exert on microorganisms. It can be observed in the appearance of a clear zone surrounding a Liberty head dime which has been pressed onto the surface of a nutrient agar plate culture of *Escherichia coli* and then incubated. Apparently the metal dissolves and diffuses through the agar sufficiently to inhibit growth of the organism for a detectable distance around the coin and a zone of a few millimeters can be measured.

This information, plus the knowledge of the inhibitory activity of antibiotics, suggests the use of the two in combination. By adding to an antibiotic small amounts of a metallic salt in a solution equal in molarity to that of the antibiotic but below the concentration known to exhibit antimicrobial activity, the ability of the metal to enhance the activity of the antibiotic at lower concentration levels can be determined.

Through the use of varying amounts of an antibiotic in a constant amount of nutrient broth, growth endpoints of a test organism can be determined by visual turbidity of the growth medium. If growth in the tubes is diminished significantly by the addition of another substance, dosage levels of the antibiotic may be lowered sufficiently to eliminate the harmful side-effects sometimes associated with their use. The search for suitable ways to enhance the activity of antibiotics at lower dosage levels is of very great concern to researchers in this field.

MATERIALS REQUIRED

The project to evaluate the antimicrobial activity of an antibiotic substance when acting alone and when acting in the presence of metallic salts in solution will require:

> 100 ml. of each stock solution:
> 0.001 M $ZnCl_2$
> 0.001 M $HgCl_2$
> 0.001 M $CuSO_4$;
> 100 ml. of a stock solution of 0.001 M bacitracin;
> 28 test tubes, each containing 10 ml. sterile nutrient broth;
> five test tubes, each containing 9 ml. sterile distilled water;
> and
> an 18 hour nutrient agar slant culture of *Staphyloccus aureus*.

An autoclave and an incubator will also be needed.

DEVELOPMENT

The tube method may be conveniently employed for the determination of antimicrobial activity of an antibiotic substance plus additives.

Four sets of seven 10 ml. nutrient broth tubes should be marked to indicate the metallic salt to be tested and the tube number in each set, and arranged, in order, in four separate test tube racks. Varying amounts of bacitracin should be added to the tubes in each set: 3.0 ml. of a stock solution (0.001M) to the first tube in each set; 1.0 ml. of stock solution to the second; 0.3 ml. of stock solution to the third; 0.1 ml. of stock solution to the fourth; 0.3 ml. of a 1:10 sterile distilled water dilution of the stock solution to the fifth; and 0.1 ml. of the 1:10 water dilution to the sixth. The seventh tube in each set will act as a control and will receive no antibiotic.

Tubes in the set designated to test the zinc chloride solution should receive amounts of the solution equal to and of the same molarity as the antibiotic previously added: 3.0 ml. of the stock solution (0.001M) of zinc chloride to the first; 1.0 ml. of stock solution to the second; 0.3 ml. of stock solution to the third; 0.1 ml. of stock solution to the fourth; 0.3 ml. of a 1:10 sterile distilled water dilution of the stock solution to the fifth; and 0.1 ml. of the 1:10 water dilution to the sixth. Number seven, the control, receives no zinc chloride solution.

Following the pattern established for the zinc chloride test, the set of tubes designated to test mercuric chloride solution should receive additions of that metallic salt, and similarly, tubes in the third set should be treated with appropriate copper sulfate solution additives. The fourth set of tubes, which tests the antibiotic alone and serves as a standard, should receive no metallic salt solution additives.

All tubes should be gently agitated to mix the contents uniformly. It is important that the antibiotic and metallic salt solution be mixed with the broth, but care should be taken not to splash the mixture or cause the cotton plugs or screw caps to become wet.

After all tubes have been prepared and their contents thoroughly mixed, each tube should be inoculated with 0.1 ml. of a sterile distilled water suspension of the 18 hour agar slant culture of *Staphylococcus aureus.* All tubes should again be agitated, this time to disperse the organisms throughout the liquid medium, and incubated at 37° C for 24 hours.

Reading the results can be accomplished easily by identification of growth of the organisms in a tube which shows turbidity, and inhibition of growth in tubes which remain clear. Results should be tabulated on an appropriately designed chart which provides for the tube number and concentration of each additive in micrograms per ml. of nutrient broth. Observations of growth should be recorded on the chart with a plus sign and inhibition of growth with a minus sign.

DISCUSSION AND INTERPRETATION

Group discussion of the results will sharpen student interest in the practicality of their project study. The significance of research as a tool for discovering ways to reduce effective dosage levels of antibiotics should be determined and both general and specific problems encountered by large dosages should be investigated.

The enhancement of antimicrobial activity of an antibiotic by metallic salt solution additives should be evaluated by comparing the organism's growth with the antibiotic alone and with the various solutions tested. "Visual turbidity" as a method of determining the lowest concentration of an antibiotic at which growth of an organism is inhibited should be discussed critically and its merits and limitations stressed. The practical value of the antimicrobial activity of certain metallic salts when used alone should also be considered. If some students have tested the metallic salt solutions alone, they should discuss the results in terms of minimal inhibitory concentration levels as compared with those observed when tested in combination with the antibiotic.

As an adjunct to the project, a simply prepared demonstration should be presented. Heavily inoculate 45° C sterile nutrient agar with *Escherichia coli* from an 18 hour agar slant culture; pour suspension into a sterile petri dish and allow to cool and solidify; lightly press onto the surface a Liberty head dime and, well-separated from it on the same surface, a "sandwich" dime; incubate at 37° C overnight and examine the following day. Students should compare the zone of inhibition surrounding the older coin with the absence of a clear zone around the later model. This observation will probably elicit some speculation and predictions concerning the removal of a bacteriostatic agent from our coins.

FURTHER STUDY

As an extension of this project, the same basic pattern for

development can be used to study the effects of vitamins, enzymes or other metabolites in combination with other antibiotics or with other test organisms. The use of antibiotics in teams, still another variation which seeks to discover greater activity at lower dosage levels, can be explored by individuals or groups as an alternate study.

The more ingenious students should be encouraged to re-design the basic project, using the agar dilution-pour plate technique, and after its completion to compare the results with those using the tube method. The more inquisitive students, questioning the advisability of using metals even at these low concentrations, should plan and conduct a project which investigates the possible injurious effects of metallic salts, measured in parts per million, on small experimental animals.

Project—ISOLATING ANTIBIOTIC-PRODUCING MICROORGANISMS FROM THE SOIL

MOTIVATION

A pinch of ordinary garden soil, when viewed microscopically, can be seen to be inhabited by a multitude of different kinds of microorganisms. Yet, despite our knowledge that the soil receives contamination from infected hosts, few if any of the micro-organisms observed in a soil sample are pathogens. This can be explained by the concept of antibiosis: resident microorganisms and foreign pathogens, competing for the limited supply of available factors in an environment incapable of supporting all, enter into an association which casts the saprophytic residents in the role of antagonists against whose effects the alien pathogens are without adequate defense. Without knowing that these organisms existed, ancient man learned to benefit from them; over 2500 years ago the Chinese recognized the healing power of some substance in the soil when they treated boils and other infections with moldy tree bark, soil, and soy bean curd; some Central American Indian tribes are believed to have treated infected wounds with soil poultices.

Today we know that in the continuing struggle for survival, disease-producing organisms introduced into the soil are destroyed by other soil microorganisms. Since Dr. Waksman's discovery of the actinomycete *Streptomyces griseus* in a sample of field soil, many microorganisms which produce antibiotic substances have been isolated from soil samples and still the search for the yet undiscovered ones continues. To date, over a million soil samples have been examined, about five hundred antibiotic-producing microorganisms have been isolated, and only about a dozen of these have been found to secrete substances suitable for use in treating human diseases.

Using simple techniques, microorganisms can be isolated from soil samples, cultured, and routinely screened for activity against test organisms. The outcome of this quest may pay a handsome dividend—the researcher may actually unearth a new and useful organism, the source of a new antibiotic.

MATERIALS REQUIRED

Very simple materials are needed to conduct a research project which seeks to discover and isolate antibiotic-producing microorganisms from the soil.

100 gram soil samples, collected in separate sterile containers, from various localities; and for each,
one flask containing 90 ml. of sterile distilled water;
five test tubes, each containing 9 ml. sterile distilled water;
seven sterile 1 ml. pipettes;
18 sterile petri dishes;
45 tubes (10 ml. each) of sterile Bacto-mycological broth;
200 ml. sterile potato dextrose agar; nutrient agar;
sterile disks of white filter paper;
a mm. ruler
an 18 hour broth culture of *Bacillus subtilis,* and
the use of an incubator and autoclave are required.

DEVELOPMENT

100 gram soil samples, collected from a depth of five to seven inches below the ground level, should be placed immediately in individual sterile containers and stored until needed for testing. A dilution series to 1:1,000,000 should then be prepared for each

soil sample: add a 10 gram soil sample to the flask containing 90 ml. of sterile distilled water, shake vigorously and allow suspension to settle for 5-10 minutes; using a sterile pipette, transfer three separate 1 ml. samples of the clear liquid of the 10^{-1} dilution to separate correspondingly labeled petri dishes and 1 ml. of the same dilution to a tube containing 9 ml. of sterile distilled water to make the next (10^{-2}) dilution, discarding the pipette without having brought it into direct contact with the water in the tube. Dilutions should be made serially to 10^{-6}, each time plating three separate 1 ml. samples of the dilution before proceeding to the next, and taking care to use correspondingly marked petri dishes for each plating and separate pipettes for each dilution in the series.

After all samples have been plated, about 12 ml. of 45° C sterile potato dextrose agar should be poured into each plate, and, with cover replaced, each plate should be rotated gently on a table top to effect an even dispersal of the sample throughout the agar medium. Care should be taken to bring about a thorough mixing by swirling the contents of the dish without splashing the cover. When the agar is cooled and solidified, plates may be taped, inverted, and placed at 32° C for an incubation period of 5-7 days, or until growth is observed on the surface.

From the numerous plates on which growth appears, those showing well-separated colonies should be selected for study; using sterile technique, each large distinctive colony should be sub-cultured in a nutrient broth tube. Incubation of the subcultures should follow, at room temperature for several days, or until growth is identified by visual turbidity of the growth medium.

To test the unfiltered broth cultures for antimicrobial activity, small disks of white blotting paper (cut with a paper punch and sterilized in a paper-wrapped glass petri dish by heating in a 350° C oven for 30 minutes) should be moistened with the subculture broth and placed firmly on the surface of a nutrient agar plate which has been freshly inoculated with an 18 hour broth culture of *Bacillus subtilis* or other test organism. After incubation of the plates, upside down at 37° C for 24 hours, an examination of the plates should be made. Clear zones of inhibition of growth of the test organism around the impregnated disks should be used to detect the degree, if any, of antimicrobial activity. These zones should be measured with a mm. ruler and recorded for each microorganism isolated from each soil sample.

DISCUSSION AND INTERPRETATION

Students should be encouraged to compare and evaluate the results of their project work. The sources of the soil samples found to yield microorganisms having antimicrobial activity should be discussed, and if possible the organisms should be identified. The special advantages to be gained from using plates showing well-separated colonies for sub-culturing should be determined, and the techniques for isolating pure cultures from a mixture should be researched and demonstrated for application to cases where discrete colony growth was not obtainable immediately from sample dilution platings.

Students should explore the methods for determining if an antibiotic substance isolated in this project is a new one or one already known to man, and what additional factors must be considered before it may be considered useful as a therapeutic agent. In this relation they may gain some perspective of the time involved between the discovery and marketing of a new antibiotic. They may also consider reasons why, despite the existence of antibiotics known to be effective against some pathogens, additional ones are being sought, and how the reduction of the use of antibiotics in animal feeds might possibly cut down the drug resistance of certain pathogens.

How soil microorganisms produce useful substances while carrying on their life processes should be reviewed and the viewpoint that all groups of microorganisms must be regarded as potential sources of antibiotics is particularly worthy of consideration.

FURTHER STUDY

Students who have isolated an antibiotic substance should extend their study by comparing the antimicrobial activity of the inhibitory substance from the soil organism with that of commercially available antibiotic disks. Others might explore the effects of various concentrations of germicides, antiseptics and other chemical substances on pathogens and compare these with the antimicrobial effects of substances from the soil samples.

The basic project can be repeated with a soil sample shown to have organisms yielding antimicrobial activity, using additions of 0.5%–1.0% NaCl solution to the agar medium to facilitate diffusion of the substances. Other concentrations and other salts or substances such as citric acid can be added in a similar manner to determine their relative values as aids to diffusion.

More capable students, taking note that the size of the zone of inhibition is not a true indication of the degree of sensitivity of the test organism, should design and conduct a project using the tube dilution method for determination of the minimal inhibitory concentration of substances isolated from soil samples. The results of the open-ended study can then be compared with and discussed in relation to those of the basic project.

RECOMMENDED READING

Berrill, N. J. *Biology in Action.* Dodd, Mead and Company, New York, 1966.

Emerson, Ralph, "Molds and Men," *Scientific American,* January, 1952.

Klinge, Paul (ed.), *Microbiology, American Biology Teacher,* Special Issue, (August, 1968), 6, 30.

———*Microbiology in Introductory Biology, American Biology Teacher,* Special Issue, June, 1960.

Kluyver, A. J. and C. B. van Niel. *The Microbes' Contribution to Biology.* Harvard University Press, Cambridge, Mass., 1956.

Oginski, E.L. and W.W. Umbreit. *An Introduction to Bacterial Physiology.* W.H. Freeman and Company, San Francisco, Calif., 1959.

Park, J.T. and J.L. Strominger, "Mode of Action of Penicillin," *Science,* (1957) 195, 99.

Pelczar, M.J., Jr. and R.D. Reid. *Microbiology.* McGraw-Hill Book Company, New York, 1958.

Raper, K.B. "The Progress of Antibiotics," *Scientific American,* March, 1961.

Roueche, B. *Eleven Blue Men.* Little, Brown and Company, New York, 1955.

Umbreit, W.W. and E. L. Oginsky. "The Mode of Action of Antibiotics: Penicillin and Streptomycin," *Journal,* Mt. Sinai Hospital, (1952) 19, 175.

Waksman, S. *The Actinomycetes: Classification, Identification and Description of Genera and Species.* Chronica Botanica Company, Waltham, Mass., 1961.

2

POPULATION DYNAMICS

Populations are dynamic organizations which, like other biological units, have the capacity for growth, development and proliferation. Their continual fluctuations reflect the activities of individual members as they experience success in reproduction, failure to survive, and migration from one region to another.

Through an interplay of physical and biological factors the environment exerts a significant influence. Physical factors which favor life still abound as they did when life originated but they simply cannot meet the demands of all life forms and their potential progeny. From the biological sector, aging and deterioration, limitations in available energy, competition within the same species, and competition between populations of different species—earlier depicted as the four horsemen of the Apocalypse: Death, Famine, War and Pestilence—offer additional resistance. More recently, a glimpse of the role played by genetic factors has also been gained. By a system of checks and balances involving the fitness of individual members both in kind and number, each population strives to achieve and maintain a delicate balance between its enormous biotic potential and the environmental resistance to its expression.

Studies in population dynamics endeavor to discover how populations change and how, in the face of the diverse and complex system of factors which act to limit and destroy them, they may be stabilized. The implications for man are tremendous.

Project—POPULATION EXPLOSION *IN VITRO*

MOTIVATION

If the rate of human population growth were to continue to increase as it has from the dawn of history, in the year 3500 the weight of human bodies alone would equal the weight of our planet, Earth. If all of the offspring of one paramecium were to survive and reproduce, with similar results experienced by all succeeding generations, this one organism consisting of 0.000000064 cubic inches of protoplasm on January 1 would give rise to a population with a total volume greater than that of the entire universe by Christmas of the same year. These estimates represent some of the best scientific thinking concerning the threat of population explosions.

Clearly, the earth has neither the space nor the energy to accommodate all possible offspring of any one species, yet every one of our over 2,000,000 species of plants and animals today exhibits an inherent readiness to reproduce at its highest possible rate. The full expression of this biotic potential, however, cannot occur unless all environmental factors are favorable for each individual of the species to survive and reproduce. Not usually observed in nature, this phenomenon can be demonstrated in a controlled laboratory situation where the progeny of a single unicellular organism can produce, in its limited environment, a veritable population explosion.

MATERIALS REQUIRED

A project which investigates the production of a large population from a single unicellular organism can be conducted in the laboratory with materials which are relatively simple. Each student team or group will need:

a thriving Blepharisma culture,
an 18 hour broth culture of *Pseudomonas ovalis,*
12 sterile micro culture tubes,
a sterile inoculating micro pipette,

a sterile transfer micro pipette,
a sterile Syracuse watch glass,
short lengths of glass tubing for mouthpieces,
two 18 inch lengths of rubber tubing,
non-absorbent cotton,
a Styrofoam culture tube support block and
a dissecting microscope.

Bunsen burners or alcohol lamps and access to an autoclave and an incubator will also be required.

DEVELOPMENT

When the protozoan Blepharisma is cultured in a broth medium bacterized by living *Pseudomonas ovalis,* the protozoans feed on the bacteria and, if other growth conditions are also favorable, thrive and reproduce by simple fission. Micro culture tubes, each originally inoculated with a single protozoan and examined regularly twice daily over a period of 4-5 days, allow for the number of organisms present at each time period to be counted, the population explosion pattern to be established, and the development of a protozoan clone to be observed.

The use of scrupulously clean glassware is essential to the success of the project. If previously unused glassware is not available, thorough washing with Micro-Solv or other specialized tissue culture glass cleanser, three rinsings with double distilled water, and drying by evaporation will prepare available glassware satisfactorily for use.

Advance planning and preparation of equipment also play an important role. A micro culture tube can be made from a 6 cm. length of 5 mm. soft glass tubing by sealing one end by heating over a Bunsen burner flame and by flaring the other end with the handle of a file inserted in it while it is kept pliable in the flame. A non-absorbent cotton plug, inserted with forceps into the flared open end, completes the simple design of a satisfactory micro culture tube.

Micro pipettes can also be fashioned from soft glass tubing. Two transfer pipettes can be made by heating and drawing out the middle of an 18 cm. length of 5 mm. tubing, and similarly two inoculating pipettes can be prepared from a 10 cm. length. If time does not permit students to make their own pipettes, Pasteur capillary pipettes in 23 cm. and 15 cm. lengths may be used instead. In either case, all pipettes should be cotton-plugged at the large opening, using non-absorbent cotton and a dissecting needle for the purpose, and flamed to remove excess cotton extending beyond the glass edge.

All glassware—micro culture tubes (placed in small wire culture tube baskets and wrapped in brown wrapping paper), pipettes (grouped by size in cotton plugged test tubes and similarly wrapped), and Syracuse watch glasses (separately wrapped)— should be sterilized by autoclaving at 121° C for 20 minutes under 15 pounds pressure. After sterilization, glassware should be kept wrapped until needed.

One liter of culture medium will provide amply for an entire class. One or more students can be assigned to its preparation:

Mix 2 gm. of Cerophyl powder in 100 ml. of double distilled water; heat gently to boiling; stir constantly and boil for one minute.

Add 0.2 gm. $CaCO_3$ and filter.

Add double distilled water to make a total volume of one liter and mix well.

The culture medium should then be dispensed, 20 ml. per culture tube. All tubes of media, cotton-plugged, should then be autoclaved at 121° C for 20 minutes under 15 pounds pressure. If not used immediately, sterile media should be stored in a refrigerator until needed.

The day before its desired use, the prepared culture media should be bacterized, one 20 ml. tube for each student team or group. Using sterile technique to make the transfer from a nutrient agar slant stock culture, each tube of media should be inoculated with *Pseudomonas ovalis* and incubated at 37° C for 18-20 hours.

After all advance preparations have been made, students should assemble materials needed for the project, throughout which great care should be taken to practice sterile technique. A transfer pipette, removed from its sterile tube container and fitted with an 18 inch length of rubber tubing and a glass mouthpiece, should be used to transfer the 20 hour bacterized culture media from the 20 ml. supply to individual micro culture tubes. Each of 12 micro culture tubes should be filled about two-thirds full with the bacterized media. A sample of Blepharisma should then be placed in a sterile Syracuse watch glass and examined under the dissecting microscope. Using an inoculating pipette, fitted with an 18 inch length of rubber tubing and a glass mouthpiece, one organism should be removed from the culture and transferred to the bacterized media in a micro culture tube. This procedure should be repeated for each of the 12 tubes in the series, with checks made under the dissecting microscope to ascertain that there is one and only one organism in each tube. The characteristic purple color of Blepharisma will facilitate this check. All inoculated micro culture tubes should then be placed in Styrofoam blocks, cut out to support them, and incubated for the duration of the project at 25° C.

Tubes should be examined twice daily under the dissecting microscope. The number of cells present in each tube at each time period should be recorded on a suitably constructed chart which provides for the tube number, date and time of observation, and number of organisms counted. The data should also be entered on a graph which plots the number of organisms against time, with counts being taken for about 4-5 days or as long as the number of Belpharisma can be counted accurately.

DISCUSSION AND INTERPRETATION

After about the fourth day, students will probably begin to encounter difficulty in making accurate counts of the number of organisms in tubes containing thriving populations. While the reasons for this will, of course, be obvious, the situation should be compared with that in other tubes in which fewer organisms have developed during the same period of time and under the same conditions. Students should be encouraged to offer possible explanations for this difference and to assess the values of averaging the number of organisms in all tubes as a means of securing a truer picture of the nature of population growth. All factors affecting population growth should be considered and a comparison made of the laboratory study with that actually expressed in nature.

The laboratory observation of a population explosion in Blepharisma should be viewed as a representative study of the biotic potential pattern of every living species. Students should refer to literature which reports the biotic potential of Norway rats, giant puffballs, houseflies, mackerel and other organisms where the figures cited point up dramatically the enormity of the possibilities and the hopelessness of the situation, were it to occur. The work of Malthus and others should also be researched and discussed in the light of implications for man.

FURTHER STUDY

Students who question the specified temperature of 25° C for incubation of the Blepharisma should be allowed to place some of their micro cultures at 20° C, and by comparing the results obtained, determine the effect of temperature on the population growth. If available, a different protozoan such as Paramecium can be used in place of the Blepharisma in one or more series of tubes and, run in parallel with the basic project, serve as a comparison study of population growth of different species.

As an extension of this project, a population explosion study using a multicellular organism can be included. Drosophila, which is easily handled and maintained in the laboratory, is a suitable experimental organism and, if time and facilities permit, white mice usually elicit much student enthusiasm.

The project also suggests many open-ended studies to be pursued by highly motivated students. Variations of techniques can be employed for isolating and establishing pure cultures from a mixed protozoan pond water sample, and the development of a protozoan clone, demonstrated here, can be used as a point of departure for initiating cell and tissue culture studies.

Project—EFFECTS OF ENVIRONMENTAL FACTORS ON GROWTH PATTERNS OF MICROORGANISMS

MOTIVATION

Bacteria are among the most prolific of the microorganisms. When provided with optimum growth conditions, most will engage in metabolism at a very rapid rate; assimilation of nutrients from the environment causes the cells to grow and, after reaching a critical ratio between surface and volume, divide in two in order to re-establish a favorable S/V ratio. Repetition of this activity, in some cases every 20 minutes, leads to an increase in the number of cells which, in the uni-cellular forms, effects an increase in the population. Uninterrupted growth, if continued at this rate for seven hours, would see the progeny of a single bacterium reaching over a million, and it is estimated that were this to continue for hours, enough bacteria would be produced to cover the earth with a layer one foot deep. Fortunately, optimum growth conditions are present in any one growth medium for only a very limited period of time.

The growth of bacterial cells in fresh culture medium can be observed to follow a characteristic pattern consisting of:

1. the *lag phase* during which the cells are adjusting to the environment and there is no significant increase in number,
2. the *logarithmic growth phase* during which the logarithm of the number of cells increases linearly with time,
3. the *stationary phase* during which the rates of growth and death are balanced and the living population of cells reaches the maximum that the culture can support, and
4. the *logarithmic death phase* during which the logarithm of the number of viable cells in the population decreases linearly with time, although in some cases a few survivors may be present indefinitely.

The various factors of the environment exert a profound influence on the rate of bacterial growth. By varying any one of these factors, a comparison with the normal growth pattern can be made and the findings applied to the area of related problems of practical importance to man.

MATERIALS REQUIRED

Each run of this project will require:
an 18 hour trypticase soy broth culture of *Escherichia coli;*
eight tubes of sterile distilled water, 9 ml. each;
56 sterile petri dishes, eight per time period;
nine sterile 1 ml. pipettes;

one sterile 10 ml. pipette;
56 tubes, 15 ml. each, of sterile trypticase soy agar at
45° C, eight per time period;
a Quebec colony counter; and
a hand tally for each student.

An autoclave or pressure cooker, an incubator set at 37° C and a room temperature incubation zone should also be available.

If lack of time or materials prohibits individual study of the entire project, it may be conducted as a group project with individuals within each group assigned a different age culture for diluting and plating. Data collected by the entire group can then be consolidated on one composite class chart for analysis and interpretation.

DEVELOPMENT

The serial dilution-pour plate count technique can be employed to determine the effect of temperature on the growth rate of microorganisms. A series of culture samples plated at various dilution levels and incubated at one temperature is compared with a similar series incubated at a different temperature.

Fifty-six petri dishes should be appropriately labeled, by marking their covers, to provide four plates for each dilution:

10^0 and 10^{-1} at 0 hours,
10^{-1} and 10^{-2} at 2 hours,
10^{-2} and 10^{-3} at 4 hours,
10^{-4} and 10^{-5} at 6 hours,
10^{-6} and 10^{-7} at 8 hours,
10^{-7} and 10^{-8} at 18 hours, and
10^{-7} and 10^{-8} at 24 hours.

Two plates for each dilution within each incubation period should be designated as Set A, to be incubated at 22° C; the remaining plates, comprising a similar Set B, should be so marked for incubation at 37° C.

Proper arrangement and handling of the culture dilutions allows for simultaneous serial dilution and plating of the culture. To identify the culture dilutions for plating in correspondingly marked plates, the broth culture should be placed to the left of eight sterile water dilution tubes, the culture being marked 10^0

and the dilution tubes, consecutively, 10^{-1} through 10^{-8}. By gentle agitation of the broth culture, the organisms can be dispersed uniformly and, using a sterile 1 ml. pipette, the transfer of 1 ml. of the undiluted culture should be made to each of the four correspondingly marked petri dishes. A similar 1 ml. culture sample should be transferred to the water blank marked 10^{-1}, with care being taken to deliver the full measure of 1 ml. without bringing the pipette in contact with the water. After use, the pipette should be discarded.

A clean pipette should be used to thoroughly mix the 10^{-1} culture dilution. This can be accomplished by gentle agitation, drawing the dilution mixture into and expelling it from the pipette 3-5 times. Transfer of 1 ml. samples of this dilution to each of the petri dishes correspondingly marked should be made, followed by the transfer of a 1 ml. sample to the water blank marked 10^{-2} and, again, discard of the used pipette. Serial dilution of the culture should be continued through the series of water dilution tubes, each time transferring samples to correspondingly marked petri dishes before making the next dilution. Care should be taken to use sterile technique throughout the procedure, to use separate pipettes for each dilution, to deliver an accurately measured 1 ml. sample every time, and, in the interest of dispersing organisms uniformly throughout the diluent and securing a good sample for each plating, to mix each dilution thoroughly.

After all 1 ml. samples have been plated in properly marked petri dishes, 15 ml. of sterile 45° C trypticase soy agar should be poured into each dish. Gentle rotation of the dish on a table top will serve to mix the agar and sample uniformly, but care should be taken to prevent any splashing on the sides or cover of the dish. Following a 10 minute cooling period, during which the agar-culture mixture is allowed to solidify, 2 inch strips of masking tape should be used to secure tops to bottoms before the dishes are placed, upside down, in their assigned incubation zones: Set A at room temperature, Set B in a 37° C incubator.

As the desired age of the cultures is reached, the plates should be removed from their incubation zones and, if possible, colony counts taken immediately. If immediate counting is not convenient as incubation periods are terminated, plates may be removed from the incubation zones and refrigerated until all plates are ready for counting. In either case, that dilution pair from each time period which appears to have a countable number of colonies present should be selected for counting and those with uncountable numbers discarded.

By placing each countable growth plate on a Quebec colony counter, a physical count can be made of the discrete colonies that have grown from individual organisms dispersed through the agar medium during the plating procedure. These counts, conveniently recorded on a hand tally, should then be entered on an appropriately constructed chart. Colony counts for duplicate plates for each incubation temperature, dilution, and age of culture should be tabulated and, considering the average number of colonies for each situation and the dilution factor, the number of organisms per ml. of culture should be calculated. This data can be plotted on both arithmetic and semi-logarithmic graph paper by plotting, in each case, the number (or logarithm) of cells at each age of culture against the time in hours.

Using this same basic pattern for development, the effects of pH of the culture medium, light intensity and food supply on the growth rate of microorganisms can also be studied.

DISCUSSION AND INTERPRETATION

Some students may observe that, while the rate of growth of two different *Escherichia coli* populations differs widely, the total number of individuals ultimately produced in each is somewhat similar. If this occurs, discussion of the phenomenon and possible reasons to explain it should be encouraged. It is the effect of environmental factors on the *rate* of population growth with which this project is primarily concerned.

Identification of the characteristic growth curve of each population and its main phases should be accompanied by an interpretation of what is actually going on at each stage of population growth and by the development of an awareness of the similarity of pattern for all single species populations. It might be well to determine at this point which stage of population growth the human population has already passed through, which stage we are now experiencing, and what the future may hold in store.

The nature of exponential growth rates, advantages of the use of semi-logarithmic graph paper, and reasons for expressing the growth rate of a microbial culture in terms of generation time rather than in terms of an increase in the number of cells per unit of time are topics for consideration. The stage of growth most markedly affected by environmental factors should be noted and evaluated in terms of the project and its extensions.

If many aspects of the project have been studied, so too have many factors of the environment been observed to exert a limiting effect on the biotic potential of a population. These effects should be viewed as the environmental resistance to growth potential under ideal conditions, and the formula for any given population as an interaction of these two dynamic forces should be researched in the literature and reported for discussion.

The importance of plating out a dilution series as a microbiological technique should be discussed fully with attention given to the many factors which may affect the accuracy of the pour plate-count technique as a method for establishing a growth curve. Other methods, such as the use of a hemocytometer for making a direct microscopic count of organisms in a sample of known culture dilution and the use of a spectrophotometer to determine the number of organisms by transmission of light through a sample of culture, should be discussed and, if possible, demonstrated.

The information gleaned from the studies about effects of environmental factors on the growth patterns of microorganisms must go beyond the project; reading and discussion should also include applications of this information as it relates to plants and animals in which man has an interest, to the medical treatment of infectious diseases, and to the fermentation industry.

FURTHER STUDY

The availability of a spectrophotometer will permit the employment of an alternate method for the determination of a microbial growth pattern. A project in which a measure of optical density is used to determine the growth pattern of *Escherichia coli* should be run in parallel with the pour plate-count technique project so that the results obtained by the two methods can be compared.

Because of its short generation time, the yeast *Saccharomyces cerevisiae* can also be used satisfactorily for population growth studies. One or more groups of students can be assigned to study the effects of temperature on the rate of population growth, using these organisms grown at room temperature and at 12° C. Using a microscope counting chamber for determination of number of organisms in a sample, growth curves can be plotted and the data provided used in the formulation of generalizations concerning the effects of temperature and other environmental factors on growth patterns of microorganisms.

Some students will probably be concerned with the implied annihilation of microbial and other populations due to limiting factors in their environments. This curiosity can be channeled into active participation in projects which investigate the effects of specific environmental pollutants on both macro and micro organisms and how, if ever, the organisms are able to resist these adverse conditions.

Project—EFFECTS OF COMPETITION ON POPULATION DENSITY

MOTIVATION

Population counts are usually expressed as numbers of "microorganisms per milliliter," "clover plants per square meter," and "birds per hectare" because the size of a population has little meaning except in terms of space.

That a given area can supply only a certain amount of food and space is obvious. Naturally this places restrictions on the expression of the biotic potential of organisms in the area, and inevitably under these conditions there will be competition among the living things for the available food and space. The competition may be *intraspecific*, in which individuals within the same species compete for space, mates and energy, or it may be *interspecific*, involving populations of different species living in the same region in competition with each other. Whether by deliberate and aggressive acts such as those observed in a predator-prey relationship or by more subtle modes of action, sometimes through the production of metabolic substances which prove harmful to the competitor, competition has various effects on populations. It may effect a balance, the extinction of one or more species, or cyclical fluctuations in the competing populations which allow both to survive.

Faced with the threat of food and space limitations and their effects on the population of his own species, the scientist's

concern with population density studies has very real and meaningful applications to a major problem in our world of today and
tomorrow.

MATERIALS REQUIRED

The basic project for the study of competition and its effects on
population density may be conducted with:
   cultures of living organisms—one culture of each of
   the following:

   Paramecium aurelia
   Saccharomyces exiguus
   Aerobacter aerogenes;

   medium for yeast culture;
   sterile pipettes, 1 ml. and 5 ml.;
   Erlenmeyer flasks, 500 ml.;
   culture bowls;
   depression microscope slides;
   capillary pipettes;
   a hemocytometer, and
   a microscope.

For the additional suggested demonstrations and class project,
cultures of the following will also be needed:
   Didinium nasutum
   Paramecium caudatum
   Bacillus pyrocyaneus and
   Pseudomonas aeruginosa.

DEVELOPMENT

Introducing paramecia to a thriving yeast culture allows
for the study of a relationship in which predator and prey populations both survive. A decrease in the prey population is followed closely by a decrease in the population of predators and,
although temporary, this relief period suffices to enable the prey
to rebuild its dwindling population. Naturally this also renews the
resources for the predator, whose population then experiences
another increase while the prey again decreases. This phenomenon
occurs cyclically in an oscillating pattern.

Preparation. Prior to the initiation of the project, the following
preparations should be made.

To prepare a satisfactory medium for the growth of *Paramecium aurelia*, add 2 gm. of Cerophyl powder to 100 ml. of distilled water and gently heat to the boiling point. Stirring constantly, let boil for one minute, add 0.2 gm. $CaCO_3$ and filter. Add sufficient distilled water to make a total volume of one liter and autoclave at 121° C for 15 minutes under 15 pounds pressure. Allow to cool to room temperature and inoculate with a water suspension of *Aerobacter aerogenes*. Then inoculate with paramecium culture, adjust to pH 7.0 and maintain, in culture bowls, at room temperature.

A culture of the yeast *Saccharomyces exiguus* should also be made. A satisfactory yeast culture medium can be prepared by adding to 1 liter of distilled water the following:

2.0 gm. bacto-yeast extract,
3.5 gm. peptone,
2.0 gm. monobasic potassium phosphate and
30.0 gm. glucose.

The mixture should be stirred constantly while gently heated to boiling, boiled for one minute, and dispensed in 10.0 ml. quantities in culture tubes. Tubes of the medium should then be cotton-plugged and sterilized by autoclaving at 121° C for 15 minutes under 15 pounds pressure. When cool, tubes of culture medium should be used immediately or stored in a refrigerator until needed.

Developing the Project. Each student group should be supplied with 20-24 tubes of the culture medium, or one tube for each day that the project is to be run. All tubes in each series should be inoculated with one drop of a water suspension of *Saccharomyces exiguus* culture. After dispersing the organisms uniformly through the medium in the first day tube, a hemocytometer should be used to make direct microscopic counts of three separate samples taken from the tube. The use of capillary pipettes for making the transfer from tube to hemocytometer is recommended, as is the averaging of three separate counts to determine the population density of the prey organism. The first day tube should then be discarded.

The number of beginning predator organisms should also be determined. A sample of the *Paramecium aurelia* culture should be transferred to a Syracuse watch glass and examined under a dissecting microscope. As located, individual organisms can be transferred by capillary pipette to each of the remaining culture tubes in the series. Ten to twelve paramecia should be transferred to each tube, with care being taken to transfer exactly the same number of organisms to each tube and to avoid the inclusion of any excess paramecium culture medium with the transfer. The number of predators thus introduced to each tube and the number of prey organisms, previously determined by direct microscopic counting, should then be recorded on a chart which will accommodate daily counts for the number of days the project is to be run.

All tubes should be incubated at 25° C in reduced light until the next day. Using the direct count technique, the number of both predator and prey organisms should be determined for the second day culture tube, after which the tube should be discarded.

By repeating the incubation and counting procedures, predator and prey populations can be determined daily over a period of 2-3 weeks, using a previously unsampled tube for each day's examination. If and when necessary for greater ease and accuracy in counting, cultures can be diluted, serially, in distilled water before sampling. The dilution factor, of course, must be considered when determining the population density. The average of three samples should be recorded for each organism each day, after which the daily population for each should be graphed, plotting density against time. Growth patterns for each population should develop over a period of 2-3 weeks, during which time 2-3 cycles may be observed and the oscillating pattern of predator and prey populations in cyclical fluctuation established.

DISCUSSION AND INTERPRETATION

The prepared graphs should be examined and discussed. The cyclical pattern in which both predator and prey populations increase and decrease should be identified and interpreted in the light of how one affects the other. Students should be encouraged to give examples of this oscillating pattern in competing populations in nature and of the importance of cyclical fluctuations which permit two competing populations to exist in the same environment without causing the extinction of either. Examples of this delicately balanced cyclical relationship between predator and prey should be cited, and well-documented cases, such as the Kaibab Deer Incident, should be researched in the literature and reported for discussion.

Short term investigations into other modes of competition will serve to generate greater interest in the project study. Daily observations of a mixed culture of Didinium *nasatum* and *Paramecium caudatum* will reveal that an increase in the Didinium population is accompanied by a decrease in the number of Paramecia. To dramatize the method by which this population change is effected, students should place a drop of thriving Didinium culture on a depression slide and, while examining it under the microscope, add a drop of the Paramecium culture to the same slide. In a matter of minutes the Didinium may be observed to attack and engulf one Paramecium after another until all have been devoured. This identifies the Didinium as the predator and the Paramecium as its doomed prey in an irreversible competitive relationship among protozoan species. In the culture bowl the predator population may be seen to grow until the prey is exterminated, after which the predator population, its energy resources depleted, will also decline and become extinct.

Consideration should be given to the factors which might offer protective advantage to preyed-upon organisms. Built-in protection in the form of reproductive rates higher than those of the predators, an environment in which a continuing source of food is made available, and the opportunity for emigration to a region where the threat of predators is not so great are factors which should be evaluated for both open and closed systems.

The advantages to be gained from a predator-prey relationship should also be reviewed. Students will, no doubt, be aware that while some individuals perish in such an interaction, there is a long-range benefit for the population. It is the weak and sickly individuals that are usually preyed upon. Through natural selection the strong and healthy are allowed to live and reproduce. This phenomenon, discussed from the standpoint of natural selection, will probably elicit much student comment and may contribute to a better understanding of the method by which evolution progresses.

That various factors within an environment contribute to the success of a population can be illustrated by a simple demonstration which involves paramecia growing in a bacterized medium. If divided into two equal parts and examined to confirm an equal population density in each, one half of the thriving culture can be exposed to a strong light source while the remaining half is maintained in an area of reduced light. Daily observations, which reveal noticable differences in the population density of paramecia in the two cultures, will pose a problem situation about which students can speculate. Whether the strong light exerts a direct influence on the population or if the total effect is, perhaps, indirect, with its initial effect exerted on the bacteria, should be explored.

Population Succession. To engender some thinking that the dependence of living things upon each other may cause the population of a given environment to change over a period of time, the dynamics of population succession in a culture bowl can be observed. Samples of hay or lettuce leaf infusion, boiled in distilled water and left undisturbed in open culture bowls for one week, should be transferred, by capillary pipette, to a depression slide for microscopic examination. Daily observations, made under both low and high power of the microscope, will reveal a succession of several different microorganisms that inhabit the culture over a period of 4-5 weeks. The succession of several populations can be seen to follow an interesting pattern in which a simple microorganism is observed to appear, produce a thriving population and then slowly decrease in number while an organism of a different type repeats the same pattern.

Discussion of this phenomenon should recognize that a laboratory simulation of a pond water succession occurs more rapidly than in nature. However, both exhibit the same patterns in which changing environments are accompanied by changes in the community, with various populations succeeding each other as the dominant and most abundant life form. The development of this concept is basic to an understanding of ecological succession in communities and might well serve as an introduction to this broad study.

FURTHER STUDY

Students wishing to extend the investigation of the phenomenon of succession should be encouraged to design and conduct a project in which culture bowls of pond water are set up in varying conditions of light, temperature and pH. Others might research another laboratory study of succession which tests the growth of the mold *Neurospora crassa* in autoclaved Biotin-free media both with and without the previous growth of *Serratia marcescens* or other bacterial organism.

Referral to pertinent literature also provides motivation for further population density studies. After researching John Emlen's experiments with populations of house mice in some old buildings at the University of Wisconsin, some highly motivated students should adapt some of his procedures to a project which investigates the maximum size population that can be supported in a given area. The mode of achieving a population balance within the carrying capacity of the environment and the psychological effects of limiting factors of food and space should be noted, with special implications for the human population.

Another motivated group, after having investigated techniques for handling and culturing fruit flies, can plan and execute a population density project using three *Drosophila melanogaster* populations grown separately in an 8 dram vial, a 100 ml. culture jar, and a 1 liter flask, or in any three containers of different size.

The dynamics of intraspecific competition in populations is an area which is richly fertile for the development of independent student research. Some modes of action, more subtle than the well known predator-prey relationship, offer opportunities for investigation into intriguing situations. A comparative study which investigates the competitive relationship between two species of paramecia can be studied by two or more student groups. Each group, using bacterized cultures of *Paramecium caudatum* and *Paramecium aurelia* in competition, can be assigned a different bacterium to use in the growth medium. A comparison of results obtained using *Bacillus pyocyaneus*, *Aerobacter aerogenes*, *Pseudomonas ovalis* and other organisms for bacterizing the medium will prove interesting and will provide a stimulus for the determination of the role played by the environment in a competitive relationship.

A project which investigates the effects of a reported growth-inhibiting substance produced by large grassfrog tadpoles in a population which also includes smaller ones should also be planned and conducted by a small group of interested students. Similarly, the competitive advantage provided a "killer" strain of paramecia, through its production of the substance *Paramecin*, can be investigated in a culture which also includes a sensitive strain.

In this area, where students will no doubt have many additional and original ideas to pursue, they should be encouraged to do so and to report their findings to the class.

Project—GENOTYPE COMPETITION IN DROSOPHILA

MOTIVATION

Some population changes have their origin on the gene level. In most populations where there is an almost random pairing of genes during fertilization, some alleles, because of their selective advantage in a particular environment, become more abundant than others. This change in gene frequency in the total population gene pool may effect a new equilibrium which will persist until it is upset by one or more factors endowed with the capacity for altering the equilibrium anew. Studies of changes in the characteristics of a population offer a unique opportunity to gain a glimpse into one of the mechanisms of evolution.

While populations are complex units to study, techniques borrowed from population geneticists may be employed to investigate changes in the characteristics of a population in the laboratory. Drosophila, the common fruit fly, is most conveniently used for studies of genotype competition, but human populations, as well as those of wild plants and animals, are also subject to the same influence.

MATERIALS REQUIRED

A project study of genotype competition in Drosophila grown in a population cage will require:
    one population cage for each study;
    wild type Drosophila cultures;
    mutant Drosophila cultures:
        mutant *ebony;*
        mutant *white* eye;

Drosophila culture media;
Drosophila etherizers;
camel's hair brushes; and
a dissecting microscope.

DEVELOPMENT

Although changes in the frequencies of competing genotypes
are difficult to observe in nature, they can be conveniently studied
in the laboratory using Drosophila grown in a population cage. By
placing a mutant or other aberrant form in competition with the
wild type, the relative success of the competing genotypes can be
determined by an examination of the adult genotypes produced
from egg samples taken periodically from the cage. Using this
technique, the project investigates the ability of the *ebony* mutant
to maintain an appreciable frequency in a fruit fly population over
a period of time.

A simple population cage, similar in design to that used by
many investigators, can be made from a standard size (13" x 8" x
4½") polyethylene refrigerator pan with a tight-fitting transparent
lid. A four inch square of fine wire screening stapled and taped
over a carefully cut out section of the lid will provide adequate
ventilation, and a series of ten 1 1/8 inch diameter holes cut in one
side and the bottom of the pan will accommodate a like number
of one ounce wide-mouth food cups. Two holes in the side of the
pan are designated for food cups by which flies are initially
introduced to the.cage and egg samples are collected periodically;
the remaining eight, on the bottom, provide for the supply and
replenishment of food through cup replacement on a rotating
schedule.

A supply of sterile food cups should be filled to the rim with
reconstituted instant Drosophila medium or other preferred formula prepared in the laboratory. After the media has solidified,
two cups should be prepared for immediate use: a wedge of media
cut out of each and discarded, the remaining surface sprinkled
with yeast suspension, and a folded strip of sterile filter paper
inserted into the space from which the wedge was removed. The
two food cups, dated, should then be inserted into two cage
bottom holes. All other cage holes should be plugged with number
15 corks and all other food cups should be stored in a refrigerator
until needed.

Thriving cultures of the *ebony* mutant and of the *wild* type *Drosophila melanogaster,* kept available for starting the project, should be examined for genotype and sex differences and handled for practice in etherizing and transfer technique.

Mutant *ebony* flies should be lightly etherized, counted, and transferred to a clean, empty food cup. Care should be taken to count two hundred flies, with approximately equal numbers of each sex represented, and to use deft strokes of a camel's hair brush while making transfers to the food cup. The cup should then be placed in one of the holes in the side of the cage and, similarly prepared, a cup containing *wild* type flies should be inserted in the remaining side hole. If not over-etherized, flies will revive and fly into the cage. If, however, after examination of the transfer cups on the following day, dead flies are found in the cups, they should be replaced by living ones to maintain the correct number and distribution of each genotype in the beginning population. When empty, transfer cups should be removed from the holes in the side of the cage and quickly replaced with corks. To prevent loss of flies during this and similar operations, a 100 watt electric light bulb held near the lid of the cage at the opposite end will attract the flies, thus keeping them momentarily away from the open escape hatch.

It is to be expected that flies in the cage will mate at random and lay eggs in the food cups. Every third day another food cup should be entered into the feeding cycle until all corks in cage bottom holes have been replaced by food cups. Thereafter, food supplies will replace the oldest cup, as determined by dates marked on each. The feeding cycle, thus established, should continue for the duration of the project.

To determine the frequencies of the competing genotypes, samples of eggs being produced by the population must be obtained, allowed to develop in isolation, and the resulting adults examined. A processed food cup inserted and allowed to remain in one of the side holes for 24 hours will collect many eggs. Microscopic examination of this exposed food will facilitate the identification of a section on which eggs have been deposited. Thus identified, the egg-infested food should be placed in a culture jar containing Drosophila media and incubated at room temperature. As adults develop they should be removed, etherized, and counted by phenotype.

The results of each egg sample should be posted on a chart which records the number and description of different phenotypes developing from the eggs and the gene frequencies, as calculated by the Hardy Weinberg formula. The sampling and interpretation procedures should be repeated every 3 weeks over a 2-3 month period or for as long as the project is scheduled to run.

In a similar manner, the project can investigate competition between other genotypes for a comparative study.

DISCUSSION AND INTERPRETATION

Data recorded on the population gene frequency charts should be examined and compared. The relative order of fitness of the genotypes should be identified and discussed, with attention focused on possible evidence of "selection." It should also be determined if this study involving a single gene difference shows evidence of one genotype being eliminated by the other or if there appears to be a trend toward the establishment of an equilibrium. Students should be encouraged to cite specific examples of how genotype competition may bring about change or stability to a given population.

The Hardy Weinberg Principle should be researched and its practical applications illustrated. Familiarity with the use and meaning of the formula can be accomplished by the determination of the frequencies of genes for straight and curly hair, free and attached ear lobes, ability or inability to roll the tongue and other traits observable in student phenotypes. Initially personalized on a class or school level, this analysis can be expanded to include other gene frequencies in the human population.

The project findings should be evaluated in the light that a laboratory situation does not truly represent a similar situation in nature. However, the correlation between the fitness of genotypes in the laboratory and in nature should be researched and reported for discussion. Consideration should also be given to the practical aspects of population genetics studies, such as the knowledge of the frequency of genes for specific blood types, which helps to determine the quantity of these types to be kept available in blood banks for a particular human population.

FURTHER STUDY

Other competitions involving different mutants and sex linked genes vs. the *wild* type Drosophila should be considered for further investigations using population cage studies. Students may use *sepia, forked,* or *Lobe* in competition with each other.

Noting that environmental factors have an effect on a population, sometimes giving a slight advantage to one genotype over another, some students may wish to probe this aspect of genotype competition by repeating the project, or one of its variations, using two populations of fruitflies in population cages kept at different temperatures.

An extension of the project can be conducted to investigate the elimination of a phenotype from the population through mating preference. Students should design and conduct a project in two parts: mating white-eyed males with equal numbers of red- and white-eyed females in the same population cage until genetic equilibrium is reached, then introducing red-eyed males to the population cage to determine if the white eye phenotype will decrease and disappear from the population due to the preference of females for red-eyed males.

After researching the topic of interspecific competition, highly motivated students should be encouraged to design a project in which *Drosophila melanogaster* is placed in competition with another species. By engaging in such a project, students may discover which is the more powerful competitor.

A practical application of the population cage can be demonstrated by one or more advanced students who plan a project in which drosophila populations are selected for increased resistance to DDT and other insecticides, to high salt concentration and to a multitude of different food preservatives.

RECOMMENDED READING

Allee, W.C. *Cooperation Among Animals.* Schuman, New York, 1951.

Andrewartha, H.G. and L.C. Birch. *The Distribution and Abundance of Animals.* University of Chicago Press, Chicago, 1954.

Benton, A.H. and W.E. Werner. *Principles of Field Biology and Ecology.* McGraw-Hill Book Co., New York, 1958.

Browning, T.O. *Animal Populations.* Harper and Row, New York, 1963.

Carson, Rachel L. *Silent Spring.* Houghton Mifflin, Boston, 1962.

Davis, Kingsley. "Population," *Scientific American,* (1963) 209, 3.

Day, L.H. and A.T. Day. *Too Many Americans.* Dell Publishing Company, New York, 1964.

Dice, L.R. *Natural Communities*. University of Michigan Press, Ann Arbor, 1952.

Dobzhansky, Theodosius.. *Evolution, Genetics, and Man*. John Wiley and Sons, New York, 1955.

————. *Genetics and the Origin of Species*. Columbia University Press, New York, 1951.

Elton, Charles. *The Ecology of Invasions by Plants and Animals*. Methuen, London, 1958.

Hardin, Garrett. *Population, Evolution, and Birth Control*. W.H. Freeman and Company, San Francisco, 1964.

Hoyle, Fred. "Forecasting the Future," *Engineering and Science*, (June, 1956).

Kendeigh, S. Charles. *Animal Ecology*. Prentice-Hall, Englewood Cliffs, New Jersey, 1961.

Lorenz, Konrad. *On Aggression*. Harcourt, Brace and World, New York, 1966.

Malthus, Thomas R., Julian Huxley, and Frederick Osborn. *On Populations: Three Essays*, Mentor Books, New York, 1961.

Odum, E.P. *Ecology*. Holt, Rinehart and Winston, New York, 1963.

Rudd, Robert. *Pesticides and the Living Landscape*. University of Wisconsin Press, Madison, Wisconsin, 1964.

Simpson, George Gaylord. *The Meaning of Evolution*. New American Library (Mentor), New York, 1949.

Sistrom, W. *Microbial Life*. Holt, Rinehart and Winston, New York, 1962.

Slobodkin, L. Basil. *Growth and Regulation of Animal Populations*. Holt, Rinehart and Winston, New York, 1961.

Strecker, R.L. "Populations of House Mice," *Scientific American*, (1955), 193,6.

Swan, L.W. "The Ecology of the High Himalayas," *Scientific American*, (1961), 205,4.

Woodwell, F.M. "The Ecological Effects of Radiation," *Scientific American*, (1963), 208,6.

Wynne-Edwards, V.C. "Population Control in Animals," *Scientific American*, (1964), 211, 2.

3

GENETICS

Genetics is a dynamic life science; its study involves interactions with the other disciplines of biology and with those of mathematics and the physical and chemical sciences as well.

The tremendous advances that have occurred since Mendel first identified the basic patterns of inheritance have gained great momentum within the past 20 years. The unraveling of the DNA mystery, discovery of regulatory genes, identification of the molecular basis of mutation and proof of the genetic control of protein synthesis have been major breakthroughs. Additionally, the public is being kept abreast of developments concerning the role of genetics in physical and mental disorders and of the genetic consequences of radioactivity and drug usage.

Constantly, new discoveries are being made. Such is the determination of today's molecular geneticists and such is the challenge of the gene.

Project—ANALYSIS OF A MONOHYBRID CROSS

MOTIVATION

Gregor Mendel could predict the offspring resulting from various crosses of his experimental pea plants—he demonstrated that prediction was possible because inheritance resulted from the chance combination of unit factors of inheritance and that certain ratios of hereditary traits appeared in the offspring produced by each type of cross. By tracing the appearance of genetic factors as they were transmitted from parents to offspring, Mendel formulated his "Principles of Heredity" and, in countless studies since, his principles have been shown to apply to other organisms as well.

Often chance produces occurrences which do not conform precisely with the expectations and it becomes necessary to determine whether the observed data are in agreement with the proposed hypothesis. By referring to a probability chart for calculated chi square values and given numbers of classes, one can ascertain how great the discrepancies between expected and observed results may be, and still be considered to be due to chance alone. Borrowed from statistical methods, this test for "goodness of fit" is invaluable when interpreting data from genetics studies.

Using *Drosophila melanogaster* as an experimental organism, students of genetics are afforded an opportunity to predict the outcome of a particular cross, to perform the cross, to examine and count the offspring produced in the F_1 and F_2 generations, and to analyze and interpret the data to see if the results actually witnessed support the prediction made or if they have been influenced by other factors.

MATERIALS REQUIRED

In a project designed to allow students to apply and interpret genetic theory, *Drosophila melanogaster* serves as a suitable experimental organism. Beginning with a monohybrid cross and expanding to include other analyses also, the project requires easily obtained materials. Each student group will need:

Drosophila cultures—one thriving culture of each—
    wild type
    ebony mutant (black body)
    Lobe mutant (reduced eye shape)
    white mutant (white eye)
    Bar mutant (narrow eye);
Drosophila culture medium;
three sterile culture jars for each cross made;

a Drosophila etherizer;

a camel's hair brush;

a re-etherizer;

labels;

a mineral oil-filled morgue jar;

sterile gauze-wrapped cotton or plastic culture jar plugs;

ether; and

a dissecting micrscope.

An autoclave or pressure cooker should also be available and, for additional crosses, cultures of each of the following Drosophila types will be needed:

dumpy mutant (reduced wings)

scarlet mutant (orange eye color)

sepia mutant (brownish-black eye color)

brown mutant (brown eye color)

vestigial mutant (vestigial wings)

white-eosin miniature mutant (pink eye color, miniature wings)

black vestigial mutant (black body, vestigial wings)

For extended studies, these additional materials will be needed:

green-albino corn seeds,

green-yellow hybrid soy bean seeds,

Mormoniella vitripennis (parasitic white wasp) pupae,

Sacrophaga bullata (black fly) puparia,

Simulium vittatum (black fly) larvae,

culture vials,

dissecting instruments, and

materials for preparing and staining tissue for microscope viewing.

DEVELOPMENT

Drosophila melanogaster provides excellent living material for the study of classic genetic crosses. Initially, students should become familiar with the proper techniques for handling and culturing the common fruit fly. The effective use of a Drosophila etherizer and subsequent handling and transfer of etherized flies with deft strokes of a camel's hair brush should be practiced until mastered, and viewing with a dissecting microscope should be

employed as a means of developing skills for identifying phenotypes to be studied and for quickly and accurately sexing the flies
by one or more of their distinguishing features.

Preparation of Media. Suitable culture media can be prepared in
the laboratory but, because moisture which tends to immobilize
flies accumulates on the surface when it is stored, quantities should
not be made in excess of the needs for any two week period. The
formula may be cut for small group use or followed, unaltered, for
a batch of about one liter: Combine 115 cc. unsulfured molasses
and 775 cc. distilled water in a large beaker and heat gently until
the mixture comes to a boil; while stirring, add 7 gm. table salt and
103 gm. cream-of-wheat cereal and continue heating for 5-10
minutes, or until the mixture is smooth and viscous; remove beaker
from heat source and add 8 cc. Tegosept M mold inhibitor solution (1 gm. Tegosept M in 10 cc. 95% ethyl alcohol); mix well and
pour to a depth of one inch into sterile culture jars; plug jars with
sterile gauze-wrapped cotton; allow to cool and store in a refrigerator until needed. Just prior to use, the surface of the media should
be sprinkled with dry yeast.

As a time saver, Carolina Instant Drosophila Medium is superb.
In dehydrated form, complete with mold inhibitor, this preparation needs only to be introduced into a sterile culture jar with
plastic foam plug, reconstituted with equal volumes of tap water,
sprinkled lightly with dry yeast and put into immediate use.

Developing the Project. Several monohybrid crosses can be made
simultaneously by student groups or may be assigned singly to
individuals or groups within a class so that a sample of each type
is studied. Students should research the genetic make-up of the
Drosophila phenotypes assigned and, on the basis of the knowledge
of genetic principles, predict the outcome for each cross to be
made. These predictions should be recorded on a chart which will
also accomodate the actual results, when they become available.

Cross I— *ebony* mutant x *wild* type

To insure that virgin females will be secured, adults from an
ebony mutant culture should be removed from the culture jar and
discarded by transfer to the morgue jar. Then, within eight hours,
newly-hatched flies should be removed and lightly etherized. From
these, six females, presumed to be virgins, should be selected and

transferred to a fresh culture jar. Careful handling while making
the transfer and maintaining the new culture jar on its side until
the anesthetized flies have revived will prevent damage to the
specimens or their immobilization due to becoming stuck on the
surface of the moist medium. To complete the cross, flies from a
wild type culture should similarly lightly etherized and six males
selected for transfer to the new culture jar. This jar should be
properly labeled, dated, and placed in reduced light   at 20° C.
Simultaneously, a reciprocal cross (six *wild* type virgin females and
six *ebony* mutant males) should be similarly prepared and incubated.

After 8-10 days, when larvae can be detected in the medium,
the adults should be removed from the culture jar to prevent the
occurrence of breeding between the two generations. About 14
days later, when the life cycle has been completed, adult flies of
the F_1 generation should begin to appear, whereupon they should
be removed, counted by sex and phenotype, and deposited in the
morgue jar. If flies begin to revive during the counting procedure,
a re-etherizer, simply constructed by placing an ether saturated
band-aid on the inside cover of a petri dish, can be placed over the
flies, as needed.

In addition to repeating the first day counting procedure, on
the second day six males and six females of the F_1 generation
should be crossed in a fresh, labeled culture jar and the procedure
for culturing the F_2 generation patterned after that established for
the F_1. The counting and recording procedures for the F_1
generation should continue through the eighth day, after which
there is a real danger that flies from the next generation might
begin to emerge and be included with the F_1. The saily counts,
thus recorded, should be totalled and compared with the expected
results as indicated by the predictions that were made. Similarly,
the adults produced in the F_2 generation should be examined and
records of observed results compared with those predicted.

Following the pattern established for Cross I, reciprocal crosses
should be made, through an F_2 generation, for

Cross II—*Lobe* mutant x *wild* type,
Cross III—*white* mutant x *wild* type, and
Cross IV—*Bar* mutant x *wild* type.

After all results have been tabulated, students should compare the expected with the observed results and, for each cross made, calculate the chi square value using the formula

$$X^2 = S\frac{(d)^2}{e}$$

Using the X^2 value and number of classes produced in each cross, the probability that a deviation as great or greater than that obtained will occur by chance alone can be read on a statistical probability chart. The X^2 values and probability of "goodness of fit" results should then be recorded on the data chart.

DISCUSSION AND INTERPRETATION

The suitability of Drosophila as an experimental organism for genetics studies should be carefully considered. Its convenient size, ability to produce many generations in a short time, adaptability to laboratory culture, and its highly prolific nature are all features which should be noted and identified as requisites for organisms selected for study.

The significance of information gained from these studies also should be noted. How an understanding of the manner in which traits are inherited has enabled geneticists to predict the type of offspring which will result from certain crosses is of considerable importance. Similarly, applications to human inheritance should be discussed. Patterns involved in the inheritance of hair color, color-blindness, hemophilia and other characteristics should be identified and explained in the light of similar patterns established in the Drosophila inheritance studies.

The project itself should be evaluated in terms of the Principle of Segregation and the techniques employed should be discussed. The importance of making crosses reciprocally and the necessity of securing virgin females for producing an F_1 but not an F_2 generation should receive careful consideration. Data showing modifications of Mendelian ratios should be analyzed and, after investigating reasons why the "goodness of fit" tests cannot readily be applied to cases involving sex linkage, interpreted for the crosses exhibiting these patterns.

The concept of mathematics as a tool for the researcher should be carefully developed. As a demonstration, while the basic project is in progress, a sample study in which 100 green-albino genetics corn seeds are germinated will help to clarify a basic principle and to dispel student fears sometimes associated with mathematical usage. The expected 3:1 ratio in the demonstration can be compared with the actual physical count of seedlings appearing with green and with white leaves. Deviations between the observed and expected results for each phenotype should be individually squared and divided by the expected number for each color and these values summed by the formula for X^2. Simple checking of a probability table for the calculated chi square value with, in this case, one degree of freedom, will reveal the probability that deviations as great or greater than those exhibited in the demonstration may be expected to occur due to chance alone. This will serve as a model of how to calculate X^2 and interpret the project data. In relation to the demonstration, the lethality of a homozygous albino condition in corn seedlings should be noted and discussed.

Students should become thoroughly acquainted with the history of genetics and its leading contributors. The works of Mendel, Morgan, DeVries, Bridges, Haldane and others should be researched in the literature, discussed, and evaluated in terms of contributions made to modern genetics.

FURTHER STUDY

Observation of the stages in the Drosophila life cycle and preparation of microscope slides of salivary gland cell chromosomes are activities which can be engaged in while the project is in session. After dissecting the salivary glands from larvae just prior to their pupation, students should use one of the stain-squash techniques developed for this type of slide preparation. Slides should then be examined under high power magnification of the compound microscope for chromosome identification and for banding arrangement differences between male and female specimens and between different genotypes. Similarly, salivary gland chromosomes from larvae of the black fly, *Simulium vittatum*, can be prepared for microscopic study and for comparison with those of the Drosophila.

As an additional investigation into the use of mathematics as a tool for simplifying the interpretation of data secured from genetics studies, a group of students can be assigned to predict the phenotypic (normal green, light green and yellow) ratio of off-spring produced by the F_2 generation of green and yellow soy beans, germinate 100 hybrid seeds, and, using the chi square test, check the goodness of fit of the actual results obtained from the 100 seed sample.

As a variation, and as a supplement to the basic project, other monohybrid crosses should be run in parallel or in place of some of the crosses specified in the project. Some student groups can be assigned to make reciprocal crosses, through an F_2 generation, using

scarlet mutant x *wild* type,

dumpy mutant x *wild* type,

or other mutant stock crossed with the *wild* type flies. The results obtained should be compared with those of the same pattern in the basic project study.

Some students will be capable of investigating whether two or more factors involved in a single set of crosses also operate in predictable fashion. These students, individually or in small groups, should be encouraged to plan and conduct a project involving a dihybrid cross in which reciprocal crosses are made between *vestigial* and *ebony* mutant stocks. Intracrosses in each of the reciprocal lines appearing in the F_1 should be made to obtain an F_2, and the chi square test should be applied to the results. Other students may follow a similar pattern of investigation using *ebony* and *dumpy* or *dumpy* and *sepia* stocks for initiating the study.

More advanced students should be encouraged to investigate the phenomena of linkage groups and cross-overs. Making reciprocal crosses between *black vestigial* mutants and *wild* type stock to obtain an F_1 and backcrossing virgin F_1 females to males of the parent stock to obtain an F_2 should be suggested for the former; and crossing *white-eosin miniature* males with *Bar-eyed* females will serve to initiate the latter. Additionally, investigation into the phenotypes produced in the F_2 of a cross between *brown*-and *scarlet*-eyed mutant stocks will give some insight into the phenomenon of gene interaction.

Unusually bright and resourceful students will find the analysis of unknown Drosophila genotypes a challenging activity. Using prescribed mating schemes as a guide, they should look for anomalies appearing in the F_1 and/or F_2 generations as a means of identifying an "unknown" mutant stock.

Genetics studies of the parasitic white wasp, *Mormoniella vitripennis,* should be considered by highly motivated students. They should design a project which studies this organism cultured in parasitized *Sarcophaga bullata* puparia and, after engaging in a one factor cross for determination of simple eye color inheritance patterns, investigate more complex situations such as a cross involving epistasis and an investigation into the phenomena of sterility and lethality associated with many of the alleles for eye color in this organism.

Students wishing to extend their genetics studies to include mammals will find mice convenient and well-suited for laboratory study. Using colored and albino mice, reciprocal crosses should be carried via intracrosses and backcrosses, to an F_2 generation with the obtained results interpreted through an application of the chi square test for goodness of fit. Other students may be assigned to demonstrate Principles of Mendelian Inheritance in a similar manner, using golden hamsters as the experimental organisms for a very rewarding activity.

Project—ANALYSIS OF FLUORESCENT EYE PIGMENTS IN DROSOPHILA

MOTIVATION

In Drosophila, the fluorescent eye pigment substances are not usually discernible as distinct compounds. However, they can be separated by the simple analytical method of paper chromatography. Widely used for the separation of organic substances such as amino acids and chlorophyll pigments, this technique when applied to Drosophila is a boon to the identification of genetic types. By capillary action, molecules of a solvent travel upward along a filter paper strip through the crushed eye of a fruitfly, separating its biochemical components into pigment spots which can be identified when viewed under ultra violet light. These compounds, called pteridines, were shown by both Ernst Hadorn of the University of Zurich and Herschel Mitchell of the California Institute of Technology to be gene-controlled.

A linear chromatogram produced by fractionating a *wild* type

Drosophila is characterized by the presence of seven pteridines—
drosopterin, isoxanthopterin, xanthopterin, sepiapterin, 2-amino-
4-hydroxypteridin, biopterin and isosepiapterin—which appear,
respectively, as spots emitting orange, purple, blue, yellow, blue,
light blue and yellow fluorescence. The presence or absence of
these pteridines and the degree to which those present are
concentrated will reveal the genetic make-up necessary to produce
other phenotypes and will give some insight into the phenomenon
of mutation and gene action.

MATERIALS REQUIRED

The basic project which investigates the chemical constituents
in Drosophila eye color mutants and compares them with the *wild*
type eye strain will require:

a Drosophila *wild* type culture;

Drosophila eye color mutants:

white, sepia, vermillion, rosy, lozenge, brown, garnet,
plum, scarlet, white-eosin, white-apricot;

large Mason jars with tight fitting covers;

Whatman #1 filter paper sheets;

glass stirring rods;

an ultraviolet lamp; and

a fractionating solvent mixture of

200 ml. n-propanol and

100 ml. 1% ammonia

DEVELOPMENT

Paper chromatography, one of the most advanced of modern
analytical techniques, can be applied readily to Drosophila genetics. A chromatogram of fluorescent pigments from the crushed
eye of a fruit fly, divided into groups and appearing as separate
spots, can be used to identify the genotype of the fly.

Several preliminary preparations should be made prior to
beginning the project:

1. The fractionating solvent mixture should be placed to a
   depth of ½ inch in the Mason jars the day before use in
   order to attain equilibrium saturation in the jar atmosphere. In the interest of maintaining constant vapor pres-

sure and insuring uniformity in results, the jars should be tightly covered at all times.

2. Large sheets of Whatman #1 filter paper should be measured to fit into the jars and on each a pencil line should be drawn 1½ inches from the lower edge. A series of small circles should be marked, with pencil, at two inch intervals along this line to indicate the locations where the eyes to be analyzed should be placed. Then, using a heavy paper or cardboard template to insure proper and uniform spacing, areas between the circles should be marked and cut out ¾ inch below the line, leaving a ½ inch strip immediately below each circle to act as a wick when the paper is introduced to the solvent in the jar.

3. Flies to be used as experimental subjects should be etherized to death ½ hour before use and maintained at room temperature.

Three etherized *wild* type fruit flies of the same age and sex should be selected and, while examining under the dissecting microscope, carefully decapitated. One at a time the three heads should be transferred to a circle on the filter paper and, with a glass rod which has been rinsed in the propanol-ammonia solvent before each application, crushed onto the paper. Using lead pencil, the specimen should be labeled to indicate its genotype.

Similarly, each of several mutant eye color forms should be processed and labeled for identification. After specimens have dried, the upper edge of the filter paper should be folded over a piece of string and stapled into position. By curving the paper to fit the contour of the jar, the paper should be introduced to the jar by hanging it from the string so that ¼ inch of the strips at the lower edge dip evenly into the solvent mixture. This will position the paper straight downward so that it does not touch the sides of the jar and place the dried specimen spots uniformly above the level of the solvent. When so organized, the jar top should be tightly screwed into position and the entire assembly placed in a darkened room or closet for 6-8 hours or until the solvent has migrated to within one inch of the top of the paper. The filter paper should then be removed from the jar and allowed to dry by hanging overnight in the dark. When dry, the chromatogram should be observed in ultraviolet light where pencil outlines should be drawn around each fluorescent spot and notes made of quantitative results before fading ensues. For each pteridine separated, its flow rate should be calculated by the formula

$$Rf = \frac{\text{distance pteridine spot travels from origin}}{\text{distance solvent front travels from origin.}}$$

DISCUSSION AND INTERPRETATION

The chromatograms should be analyzed and the pteridines seperated from mutant eye color strains compared with those obtained from Drosophila *wild* type specimens. The chemical components needed to produce each eye color should be determined and both qualitative and quantitative aspects of the analysis should be discussed.

A comparison of chromatograms prepared by different students will, no doubt, show some variation for the same phenotypes analyzed. The variables of orientation of paper, length of paper, composition of solvent, kind of paper used, concentration of substance being analyzed, kind and amount of impurities present in both solvent and solute, initial distance of solvent from the starting line, temperature, light and time should be considered as factors that make difficult an exact agreement of the same genotype analysis.

The technique of paper chromatography should be discussed. The principle that each constituent of a mixture migrates at a characteristic rate with respect to the rate at which the solvent front advances should be identified as the basis for the technique, and the growing popularity of the technique for simple analyses should be recognized. By comparing the time and skill factors involved in a more traditional method of chemical analysis involving analytical chemistry methods for separating these pigments with the methods employed in the project, students should evaluate paper chromatography as an analytical technique in today's research.

FURTHER STUDY

Many variations of the basic project can be employed to add depth to the study. Some students should be encouraged to excise eyes only and compare the chromatograms prepared with those made by using whole heads from the same phenotype. Others might try a comparison of males and females by making chromatograms of the same mutant eye color strain for each of the sexes, a comparison of older and newly emerged flies of the same genotype, or of flies of the same genotype grown at different temperatures.

In an extended study, some students should compare the body color of the *wild* type fruit fly with that of body color mutants. Brighter students should design and execute a project which investigates the pigments needed to produce certain body colors, using Drosophila *wild* type and *ebony* and *yellow* mutants as the experimental organisms.

Highly motivated students should be encouraged to check the effects of food and chemicals on pteridine production by comparing chromatograms made from flies of the *wild* type cultured in different media, with and without a chemical mold inhibitor.

Students may also extend their knowledge of the technique of paper chromatography by exploring the circular form. A chromatogram prepared by this method can serve as a comparison with the linear form prepared in the project. A further extension of the linear technique should also be investigated and interested students should plan and conduct a project in which they prepare a two dimensional chromatogram of one or more of the Drosophila genotypes and demonstrate the technique to the class.

The technique of paper chromatography can be applied to many forms for genetic analyses. An investigation into the plant world to determine the molecular structure and genetic origins of flower petals is one with great esthetic appeal. Using *Impatiens balsamina* as the experimental organism, students should employ paper chromatographic techniques for the separation of anthocyanidin and flavonol pigments from its colorful petals and compare their presence or absence in the production of a given color.

Project—GENE EFFECTS IN CHEMICAL PATHWAYS LEADING TO STARCH SYNTHESIS

MOTIVATION

Modern genetics is concerned with the role of genes as modifiers of the biochemical responses of organisms. It appears that genes, functioning as regulators of enzyme actions, control biochemical reactions which may exert profound influences on the phenotype of the individual. For example, in man the absence of the enzyme that normally converts phenylalanine to tyrosine has been found to be due to a recessive gene. Consequently, in individuals who lack this enzyme, phenylalanine accumulates and causes brain damage and severe mental retardation. Fortunately, if this condition, known as PKU, is detected in early infancy, it can be treated before brain damage occurs.

Applying this modern concept to Mendel's early genetic studies with pea plants, it may be hypothesized that genes control the production of enzymes which regulate metabolism and, ultimately, produce the morphological differences between smooth and wrinkled pea seeds. In effect, the phenotype of an individual is viewed as a reflection of its biochemical composition and that which is inherited as a developmental mechanism.

MATERIALS REQUIRED

Each student, or group, engaging in a project which explores the biochemical effects associated with a gene change in peas will need:

10 gm. smooth, spherical pea seeds;
10 gm. wrinkled pea seeds;
glucose agar;
distilled water;
Lugol's iodine solution;
filter paper and
a petri dish.

The use of a microscope, a centrifuge, and a Waring blendor or a mortar and pestle will also be required.

DEVELOPMENT

In a process common to all green plants, glucose, which has been synthesized by the process of photosynthesis, is enzymatically converted to starch. An investigation of this conversion reaction can be made to determine if differences in biochemical activity are associated with differences in pea seed phenotypes.

An enzyme extract from wrinkled pea seeds should be prepared: 10 gm. of wrinkled pea seeds, ground in a Waring blender or crushed with a mortar and pestle, should be mixed with 10 ml. of distilled water and centrifuged for five minutes. The supernatant (or clear filtrate, if centrifugation is not convenient) should be properly labeled and reserved for later use. Meanwhile, the same procedure should be followed for the preparation of an extract from the smooth pea seeds.

Glucose agar plates should be prepared for use: a mixture of 2.0 gm. of agar and 0.5 gm. of glucose-1-phosphate should be added to 100 ml. of distilled water, heated to boiling, boiled for one minute, and dispensed in petri dishes with bottoms divided into two compartments. After the plates have solidified, each should be labeled on the under surface with the letters "W" and "S" to indicate which half is to receive the corresponding pea seed extract.

Using a clean dropper, samples of the wrinkled pea seed extract should be placed within the area bearing the "W" in three well-separated positions on the surface of the agar plate. Similarly, three separate drops of the smooth pea seed extract should be introduced to the other half of the same plate. Allowing an interval of 15 minutes for the enzyme to act on the glucose, one drop of Lugol's iodine solution should be added to one of the spots where wrinkled pea seed extract had been placed and, in a like manner, one of the areas of smooth pea seed extract should be tested. After one minute, small pieces of filter paper should be used to absorb the iodine-enzyme extract mixture from the agar surface to allow an examination of the glucose agar both at and below its surface. This procedure should be repeated at 30 and 60 minute time intervals, using the remaining two sets of seed extract spots for the tests. Observations of reactions for each type of seed extract at each time period should be recorded on a suitably prepared chart which provides for quantitative as well as qualitative readings of the reactions noted.

DISCUSSION AND INTERPRETATION

From the data collected, the type of pea seed which synthesizes starch more rapidly should be determined and the effect of genes on the chemical pathways leading to this progressive synthesis assessed. The starch-forming enzyme, phosphorylase, should be researched, reported, and discussed in the light of differential chemical activity observed in the wrinkled and smooth seed extracts.

The use of iodine as a test reagent and the employment of centrifugation as a technique for the complete removal of all traces of starch grains from the enzyme extracts should be evaluated as they contribute to the project.

Morphological differences between the seed types can be demonstrated by separately preparing smear slides of cut surface scrapings from each seed type on a drop of distilled water and viewing them through a microscope for a comparison of their starch grains and leucoplasts. Physiological differences can be noted in a comparison of the uptake of water by a given number of seeds of each type when soaked overnight in separate jars containing distilled water.

Human diseases and disorders such as phenylketonuria, galactosemia, alkaptonuria and cretinism should be researched and reported to the class for discussion from the point of view expressed by the statement that "although nothing can be done about the genetic make-up of an individual, it is possible to treat the secondary effects."

FURTHER STUDY

If time permits, students may be assigned to conduct a project which seeks to distinguish the biochemical differences in three separate Drosophila genotypes. Using the *wild* type, *ebony* mutants, and heterozygotes produced by these two pure strains, flies should be selected as viable in the larval stage and transferred separately by genotype to culture jars containing minute quantities (ranging from 0.01% to 0.10%) of phenylthiocarbamide mixed with the culture media. The relationship of genotype to biochemical response to PTC, in this case an inhibitor of the enzyme dopa oxidase, should be measured by the survival rate of each genotype in each concentration level of PTC. Students wishing to extend this investigation further can explore the ability of other chemicals to "selectively" kill fruitflies, by adapting this procedure to accommodate the testing of small percentage levels of formaldehyde, paradichlorobenzene, dichlorodiphenyltrichloroethane, and other chemical substances.

Having researched the works of Beadle and Tatum, one or more of the advanced or highly motivated students might develop a project which investigates the biochemical genetics of *Neurospora crassa* using the *wild* type and selected mutant strains for the study.

Students whose curiosity is piqued by the knowledge that dwarf mice, produced by a recessive gene, have abnormally developed pituitary glands, may wish to perform a rather delicate organ transplant but lack the necessary skills to do so. Often a visit to a research laboratory can be arranged where they can observe a professional transplantation of the pituitary gland from a normal mouse into the body of a genetically dwarfed individual. The mouse, recovered from the operation, can be returned to the classroom where over a period of time it can be observed until a normal sized mouse results, even though all of its body cells, except those in the transplanted pituitary, are homozygous for the dwarf gene.

Project—MUTATION RATE IN ESCHERICHIA COLI

MOTIVATION

Mutations are believed to occur in all organisms, from bacteria to man. Generally defined as an abrupt, stable change in a gene, a mutation results from some alteration of the nucleotide sequence in the DNA molecule which brings about a slightly different biochemical effect on the cell. As such, mutations represent random changes in the controlling mechanism of all living things.

Most mutations are considered to be harmful since they bring about changes in an organism which, as a product of natural selection, is already successful and well-established in its environment. In Drosophila studies 80% of all mutations observed have been determined to have detrimental influences on the fly and almost 20% are lethal. Yet, without mutations there would be no beneficial changes to continue to provide a basis for natural selection in a changing environment. To survive, a species must maintain a delicate balance between the harmful effects brought about by too many mutations and the occurrence of some mutations which will ensure its existence.

Although harmful and deleterious in most of their effects, mutations are necessary to the maintenance of our biotic world; without them evolution could not occur.

MATERIALS REQUIRED

A project which investigates the mutation rate in *Escherichia coli* requires that sterile technique be employed throughout the experimental procedures and that each student group be supplied with:

> four 18 hour nutrient broth cultures of *Escherichia coli lac⁻;*
> 16 Endo agar plates;
> 24 9 ml. distilled water blanks;
> 28 sterile 1 ml. pipettes;
> an inoculating loop;
> a Bunsen burner or alcohol lamp;
> 0.01% formaldehyde solution;
> an ultraviolet lamp;
> a Quebec colony counter;
> a hand tally;
> lactose broth (Difco);
> phenol red indicator solution;
> 16 Durham fermentation tubes, and
> safety goggles.

An incubator should be available for use and, for extended studies, an assortment of chemicals and access to the use of X-ray equipment.

DEVELOPMENT

The wild type *Escherichia coli,* designated as *lac⁺*, ferments lactose and appears as a red colony growth when cultured on Endo agar, whereas the mutant *lac⁻* strain, unable to ferment this sugar, will appear as white colonies when grown on the same medium. Sometimes *Escherichia coli lac⁻* organisms mutate back to the wild form and can be detected by their red colony growth.

Using sterile technique and separate pipettes for each dilution, 1 ml. of an 18 hour broth culture of a *lac⁻* strain of *Escherichia coli* should be serially diluted through six 9 ml. sterile distilled water blanks to reach a 10^{-6} dilution of the original. A clean pipette should be used to transfer a 0.5 ml. sample of this final culture dilution to the surface of an Endo agar plate and the inoculum spread with a sterile inoculating loop over the entire surface by rotation of the plate 60°, twice, between three separate spreadings in order to insure its uniform distribution. The inoculated agar plate should be labeled, taped with two 2 inch strips of masking tape to secure its top to bottom, and incubated in an inverted position at 37° C for 24 hours.

The following day a physical count of the resulting red, white and red-specked white colonies should be taken, using a Quebec colony counter and hand tally to facilitate the procedure. The rate of back mutation should be calculated and recorded on a suitably prepared chart.

As a test for stability, samples of each of the three types of colonies developed should be streaked onto the surface of a fresh Endo agar plate and, after incubation at 37° C, examined with a Quebec colony counter to determine the stability of the mutations observed in their original form.

To test the effects of ultraviolet light on the mutation rate, an 18 hour nutrient broth culture of the *lac⁻* strain of *Escherichia coli* should be exposed to ultraviolet light for 2 minutes just prior to its serial dilution. Care should be taken to protect the eyes by wearing protective goggles and by avoiding any direct viewing into the lamp light. Then, following the procedure established for the untreated culture, the mutagenic effects of the exposure and the stability of the mutants produced should be determined.

In a similar manner, the effects of heat and chemicals should be investigated. Heating a *lac⁻* broth culture to 45° C for 30 minutes and adding 1 ml. of a 1% formaldehyde solution to another will serve as preliminary treatments in a study which seeks to determine the mutation rates associated with these mutagenic agents and compare them with the rates observed in the untreated culture.

DISCUSSION AND INTERPRETATION

The importance of maintaining aseptic conditions and of practicing sterile technique throughout the project should be considered. The wearing of goggles when using the ultraviolet lamp and of not looking into the lamp light should be discussed as precautionary measures taken in this phase of the project study.

If, in the project, two mutations have been observed and tested, the degree of stability of each should be noted and possible explanations for the results offered. Factors which were observed to have influenced the mutation rate should be discussed and their long-range effects assessed for *Escherichia coli* and for other populations, including that of man.

Biochemical differences between the three types of colonies isolated can be demonstrated as they exhibit an ability to ferment lactose. In a simple procedure which can be performed by an adept student, each type of organism should be introduced to a separate Durham fermentation tube containing 10 ml. of lactose broth (Difco) to which phenol red indicator solution has been added. After an incubation period of 24-48 hours at 37° C, the tube vials can be tested for gas collection and the indicator solution for the red-to-yellow color change by which the products of carbohydrate degradation can be detected.

The importance of mutations should be discussed in detail. Although generally considered to be rare and harmful, agents which increase their rate of occurrence and the resulting effects on the evolutionary process should be carefully assessed. Breeding methods employed to develop and maintain pedigrees of poodles, mink and tropical fish should be explored and reported to the class for discussion as a practical application of studies involving mutations.

FURTHER STUDY

The initial investigation into mutation rates of the bacterium *Escherichia coli* can be extended further by making use of a wide range of temperatures, formaldehyde concentrations, and ultraviolet light exposures. Additionally, other chemical substances such as alcohol, peroxide, chloroform and caffeine can be investigated as possible mutagenic agents. With the cooperation of a local dentist or hospital staff, the effects of X radiation can be studied in this regard and the resulting mutation rates compared with those observed to have been caused by other mutagenic agents.

A project in which saline suspensions of bacteria are irradiated at various dosage levels, grown on nutrient media, and examined for the occurrence of mutations should be planned by students who can enlist the aid of a dentist or radiologist. Using *Serratia marcescens* or other pigmented bacteria as experimental organisms, variations in the pigmentation can be interpreted as an indication of mutations in the pigment-producing mechanism of the cells. A further comparison of these pigments, via the techniques of chromatographic analyses, will provide a challenging and rewarding activity for a bright student.

The influence of radiation, recognized as a primary cause of mutations, can be studied as it affects other organisms also. In an investigation of the effects of differential dosage levels of X-ray exposure on the occurrence of mutations, students can pursue genetics studies involving the germination of irradiated seeds or the morphological and physiological abnormalities in the offspring of irradiated wasps and fruit flies. In another study, gene changes which alter the mechanism for the synthesis of certain nutrients can be explored using the pink bread mold *Neurospora crassa* as the experimental organism.

Some students may find working with plant life another interesting extension of the project study. They should design a project which investigates mutation rates in treated and untreated plant cells, using some of the mutagenic agents found to be effective when applied to microorganisms. For those who practice meticulous technique, an investigation into the effects of the chemical colchicine in producing polyploid cells in Tradescantia will prove an absorbing and rewarding activity.

The adaptive potential of *Escherichia coli* should be studied in depth by the more advanced students. They should be encouraged to design and conduct another project in which the identification of mutants which possess the ability to produce a strain similar to that from which they mutated will bear evidence of genetic recombinations. For a really bright and highly motivated student, this study can be expanded with the use of selective media, to include the mapping of genetic material in the bacterial chromosomes.

RECOMMENDED READING

Auerbach, Charlotte, *Science of Genetics,* Harper and Row, New York, 1961.

——— . *Genetics in the Atomic Age.* Oliver A. Boyd, London, 1956.

Barry, J.M. *Molecular Biology: Genes and the Chemical Control of Living Cells.* Prentice-Hall, Inc., Englewood Cliffs, N.J., 1964.

Beadle, G.W. "Genes and Chemical Reactions in Neurospora," *Science,* (1959), 129, 3365.

Bearn, A.G. and J.L. German, III. "Chromosomes and Disease," *Scientific American,* (1961), 205, 5.

Bernard, Claude. *An Introduction to the Study of Experimental Medicine* (translated by Henry Copely Greene). Dover Books, New York, 1957.

Bonner, David M. *Heredity.* Prentice-Hall, Inc., Englewood Cliffs, N.J., 1961.

Casarett, Alison P. *Radiation Biology.* Prentice-Hall, Inc., Englewood Cliffs, N.J., 1968.

Clevenger, Sarah. "Flower Pigments," *Scientific American,* (1964), 210, 6.

Crick, F.H.C. *Of Molecules and Men.* University of Washington Press, Seattle, 1966.

Crow, James F. *Genetics Notes.* Burgess Publishing Company, Minneapolis, 1966.

Davidson, Eric H. "Hormones and Genes," *Scientific American,* (1965), 212, 6.

Demerec, M. and B.P. Kaufmann. *Drosophila Guide.* Carnegie Institution of Washington, Washington, D.C., 1964.

Dobzhansky, Theodosius. *Evolution, Genetics, and Man.* John Wiley and Sons, New York, 1955.

Hadorn, Ernst. "Fractionating the Fruit Fly," *Scienfific American,* (1962), 206, 4.

——— . "Transdetermination in Cells," *Scientific American,* (1968), 219, 5.

Hotchkiss, T. and E. Weiss. "Transformed Bacteria," *Scientific American,* (1956), 195, 5.

Kalmus, Hans. *Variations and Heredity.* Routledge and Kegan, Paul, Ltd., London, 1957.

Kronen, R.A. and P.A. Parsons. "The Genetic Basis of Two Melanin Inhibitors in Drosophila Melanogaster," *Nature,* (1960), 186, 411.

Mirsky, A.E. "The Chemistry of Heredity," *Scientific American,* (1953), 188, 2.

Parsons, P.A. "A Widespread Biochemical Polymorphism in Drosophila Melanogaster," *American Naturalist,* (1963), 97, 375.

Peters, J.A. (ed.) *Classic Papers in Genetics.* Prentice-Hall, Inc., Englewood Cliffs, N.J., 1959.

Snyder, C.R. and R.L. James. "Low-Activity Isotopes as Bacterial Mutagens," *The American Biology Teacher,* (1965), 27, 9.

Strong, C.L. *The Amateur Scientist* (Scientific American Book of Projects), Simon and Schuster, New York, 1960.

Taylor, J.H. (ed.) *Selected Papers in Molecular Genetics.* Academic Press, New York, 1965.

Tomasz, Alexander. "Cellular Factors in Genetic Transformation," *Scientific American,* (1969), 220, 1.

Waddington, C.H. *New Patterns in Genetics and Development.* Columbia University Press, New York, 1962.

Wollman, E.L. and F. Jacob. "Sexuality in Bacteria," *Scientific American,* (1956), 195, 1.

Zinder, N.D. "Transduction in Bacteria," *Scientific American,* (1958), 199, 5.

4

DEVELOPMENT

Nothing known to man approximates the enormous potential of a fertilized egg. As it divides and grows it undergoes a progressive development in form, structure, and biochemical activity which transforms it into an immensely complex individual endowed with the characteristics of its parents.

Notwithstanding its progressively changing characteristics, the embryo, regardless of its stage of development, is an organized system in which the changes in process at any one time are perfectly coordinated. Explaining how these events take place in an orderly manner and how we may control some of the factors which may operate to interfere with desirable developmental patterns are the challenging problems to which researchers in this field are directing their attention and energies.

Project—TEMPERATURE VS. GENE-CONTROLLED DEVELOPMENT IN DROSOPHILA

MOTIVATION

While inheritance is usually considered in terms of specific characteristics, it is only the genes that are actually inherited; the characteristics are produced by the action of environmental factors on the gene-controlled process of development. More

85

readily observed in fully developed organisms such as man, where
the gene for hereditary baldness is expressed only in the presence
of certain male hormones, and in Himalayan rabbits, where a gene
which produces white hair at temperatures above 92° F will
produce hair which is black at temperatures that are lower,
environmental factors also exert great influence during the developmental stages of most organisms. Interference from internal or
external sources may influence the chemical activity of a gene,
causing it to be more strongly activated or suppressed as it acts to
control and regulate development.

Encouraged by the successful treatment of PKU, diabetes and
pernicious anemia which follow characteristically hereditary patterns, researchers are endeavoring to locate and identify controllable factors in the environment that are directly related to other
gene expressions that plague man. Hopefully, certain structural
defects and deformities that show strong hereditary patterns will
soon be revealed to be influenced by environmental means that
can then be effectively implemented in a highly rewarding break-
through.

MATERIALS REQUIRED

A project which investigates the effect of temperature on the
development of wing length in *Drosophila melanogaster* can be
completed in one month and will require:

a Drosophila *wild* type culture;
a Drosophila *vestigial* (wings greatly reduced) mutant culture;
Drosophila culture jars;
Drosophila etherizers;
Drosophila culture medium;
camel's hair brushes;
dissecting needles;
dissecting microscopes;
ether;
temperature controlled incubation zones;
white cards;
mineral oil filled morgues; and
ocular micrometers for microscopes.

If ocular micrometers are not available, Bogusch measuring
slides or transparent mm. rulers may be substituted.

DEVELOPMENT

Before beginning the project, each student group should review the procedures and techniques for handling and culturing fruit flies. Then, flies from a stock culture of Drosophila *wild* type should be lightly etherized and examined under a dissecting microscope. Using an ocular micrometer or Bogusch measuring slide, the wing length of 100 flies should be measured and the average length computed and recorded on a chart which will also accommodate data for average wing lengths of flies developed at two different temperatures.

Male and female flies should be selected from the parent stock and transferred, six of each sex, to each of four freshly prepared Drosophila media culture jars. After they have been properly labeled, two culture jars should be incubated at 28° C, the other two at 18° C. Jars should be examined daily and, when pupae appear, the adults should be removed and discarded while the pupae continue to undergo development. Three days after the emergence of adults they should be etherized and, again, wing lengths of 100 individuals from each culture jar measured and average lengths computed. Data for each designated temperature at which flies were developed should be recorded on the previously prepared data chart.

Simultaneously, the same or different individuals or groups should follow a similar procedure, employing the *vestigial* mutant with reduced wing length as the experimental organism developed, in duplicate, at 20° C, 25° C, and 30° C. All cultures should be examined daily and, when larvae appear, parent flies removed and discarded. Larvae should be permitted to develop and, three days after their emergence from the puparia, adults from each of the incubation zones should be etherized for examination. The average wing length produced at each temperature should be determined and recorded for comparison with each other and with the average wing length exhibited by the parent stock culture. Six males and six females from the culture developed at 30° C should then be placed in each of six fresh media culture jars and again divided into three groups for development at three different incubation temperatures: 20° C, 25° C, and 30° C. Again, parent flies should be discarded after larvae appear and, three days after their emergence, adults examined for determination of wing length. Data, recorded on the chart, should be analyzed and wing length of progeny, raised at three different temperatures, compared with each other and with their parent stock.

DISCUSSION AND INTERPRETATION

The results obtained by the various groups should be compared and interpreted. Wing length should be correlated with developmental temperature and the optimum temperature for the expression of genes for each wing length determined.

The observation that *wild* type flies raised at 28° C have decidedly shorter wings than their genetically identical counterparts raised at 18° C will probably elicit some questions concerning whether they are to be considered to be mutants. This point should be discussed and an interested individual or group encouraged to design and conduct an extension of this part of the project to check out the prevailing hypothesis.

Another instance of the regulatory role played by temperature in the expression of a gene can be demonstrated in the classroom if a suitable flowering plant such as *Primula sinesis,* a variety of the chinese primrose, can be secured. This variety is homozygous for genes which produce white flowers at temperatures above 86° F, red ones at temperatures below this level. The flower color, which depends on the temperature during the formation of the flower, can therefore be determined at will, merely by changing the location of the plant during this critical period. This is a dramatic way of demonstrating that, while inheritance is usually considered in terms of specific characteristics, it is actually only the genes that are inherited; the characteristics result from the action of the environmental factors on the gene-controlled processes of development.

The effects of light on the color of grains on an ear of *sun red* corn can be demonstrated, similarly, as another example of the phenomenon of environmental control over gene expression. In this respect, the production of red pigment in the aleurone cells of corn when exposed to sunlight should be researched and reported to the class for discussion. Along this same line, the various effects of sunlight on the skin of man should be considered with

explanations offered for the ability of some fair-skinned genotypes, expressed by class members, to develop a deep tan when
exposed to abundant summer sun. It is this aspect of the project
that should be highlighted, with implications for man to search for
controllable environmental factors which are directly related to
gene expressions harmful to our physical and emotional health.

FURTHER STUDY

To check out another environmental factor and its regulatory
role in the expression of genes, students should secure and
germinate corn seeds. Grown in darkness and in light conditions,
two groups of seedlings with identical genetic make-up can then be
compared as to structural differences and physiological variations.
Other green plants can be similarly employed to demonstrate that
individuals are the product of their genetic endowment and their
environment.

As an adjunct to the basic project, the effect of environmental
factors of food and space on the development of individuals can
be investigated by a small group of students who raise fruit flies in
a small vial with a scanty food supply and compare their size with
those of identical genotype raised in a larger container with
abundant food.

Amphibian development also lends itself to the study of the
effects of environmental factors on development. Highly motivated students should be encouraged to design a project which
investigates the effect of water temperature on the rate of
metamorphosis and size of individuals produced, using tadpoles or
salamanders from the same egg clutch as the experimental organisms.

Project—EFFECT OF HORMONES ON TADPOLE DEVELOPMENT

MOTIVATION

In embryonic development there is a continuing progressive
determination of structure with time. The process, initiated by the
stimulation of an egg by a sperm, is irreversible and includes not

only those activities which give form and shape to the body and its structures but also brings about permanent modification of cells as they become differentiated and specialized. What controls this complex process is not fully understood but both intrinsic and extrinsic factors have been observed to exert great influence on its pattern.

Countless numbers of starfish, sea urchins, tropical fish, snails, brine shrimp, and various insects and birds have served as experimental subjects for developmental studies in which the normal pattern has been observed to be modified by nutrition, temperature, surgical operation, over-crowding, hormones, and other chemical substances. The larval stage, which is a part of the total developmental process of most insects and amphibians, is of particular value both from the standpoint of convenience and the adaptive nature of changes in their morphological, physiological and biochemical structure. It is through experimental studies with these lower animal forms that we gain valuable insights into the control and regulation of developmental processes in the higher forms, including man.

MATERIALS REQUIRED

A project which investigates the effect of a hormone on tadpole development will require:
>   four culture bowls or aquaria;
>   pond water or Holtfretter's solution;
>   thyroxine solutions in concentrations of
>   1:1,000,000, 1:10,000,000 and 1:100,000,000;
>   petri dishes;
>   Bogusch measuring slides or mm. graph paper;
>   distilled water;
>   weighing balances; and
>   dissecting microscopes.

DEVELOPMENT

Several preliminary preparations should be made prior to beginning the project:

1. If natural pond water cannot be secured in sufficient quantity for the project, Holtfretter's solution, a simulated pond water, can be made up according to the formula:

$$NaCl \quad 3.5 \text{ gm.}$$
$$KCl \quad 0.05 \text{ gm.}$$
$$CaCl_2 \quad 0.1 \text{ gm.}$$
$$NaHCO_3 \quad 0.2 \text{ gm.}$$

in one liter of distilled water.

2. Thyroxin solutions should be prepared: 1 mgm. of thyroxin dissolved in 1 cc. of 1% NaOH and diluted to 1 liter in distilled water will yield a 1 ppm. solution which should be refrigerated until needed. This stock solution can be used full strength or diluted further in distilled water to other desired concentrations for use in the project study.

3. Sufficient numbers of tadpoles should be secured and maintained at room temperature in large aquaria. Routines of feeding with parboiled spinach and changing of fouled water three times a week should be established and practiced throughout the project.

Each of four student groups engaging in the project should be assigned a different solution in which to raise one dozen tadpoles for a period of 3-4 weeks:

1. thyroxin solution at a concentration of 1:1,000,000 (1 ppm. stock solution);
2. thyroxin solution at a concentration of 1:10,000,000 (1 part stock solution to 9 parts distilled water);
3. thyroxin solution at a concentration of 1:100,000,000 (1 part stock solution to 99 parts distilled water); or
4. pond water or Holtfretter's solution, to serve as a control.

Students should prepare and adjust the solutions assigned to room temperature conditions in culture bowls or aquaria. After 12 healthy tadpoles of approximately the same age, size, and weight have been selected for each group, their average weight and length should be determined: they should be transferred, one at a time, to a petri dish placed over a measuring slide or mm. graph paper to measure the distance from head to tail, and on a weighing balance for determination of weight. Average lengths and weights, calculated for each group, should be recorded on a chart. Tadpoles should then be transferred to the appropriate culture bowl containing the solution to which they are to be exposed. As routines of feeding and transfer to fresh solutions are continued, weight and length measurements should be taken, averaged, and recorded for these established time intervals. This routine should be continued for a period of 3-4 weeks during which time each group should make note of developmental progress, keeping a calendar of events as the tadpoles undergo their metamorphoses. Throughout the project, two composite graphs should be prepared, with each group plotting the average weight on one, using assigned colors for identification of the solutions associated with this pattern, and the average length, similarly plotted, on the other.

DISCUSSION AND INTERPRETATION

Results obtained by each of the groups should be combined to give a total picture of the project study. An analysis of the size and weight growth patterns graphed for each different treatment should be discussed in relation to the hormone concentrations. By a comparison of these growth patterns with the controls, where a normal pattern is exhibited, the effect of the thyroid hormone should be established. Predictions of what might be expected at concentrations above and below those used in the project should be encouraged as well as plans for a new project for testing these hypotheses.

The phenomenon of metamorphosis should be considered as a period of adaptive changes involving the preparation of an organism for a transition from one environment to another which is entirely different. Via group reporting on the time table of events observed in the particular situation investigated, differences in times at which there was a rapid increase in hind limb development, fore limb emergence, rapid resorption of the tail tissue, and changes in the form of the head can be directed to point up the role of the thyroid hormone in controlling the rate of these metamorphic changes.

Wigglesworth's studies of the *juvenile hormone* in certain moths usually stimulates great interest in this substance. Outside reading about his experiments with this hormone that prevents the pupation of moth larvae until they have reached their full growth and that has been observed to produce both dwarf and giant adult insects should be reported to the class for discussion.

FURTHER STUDY

The project lends itself to many variations in which another organism, such as the salamander, or other test agents, such as iodine or triiodothyronine, are used.

Students wishing to investigate the effect of a thyroid inhibitor on frog metamorphosis can engage in a project study in which tadpoles are exposed to various concentrations of thiourea solutions and compared with controls raised in pond water. Others, wishing to explore an application of this study to higher organisms, should investigate techniques for injecting embryonated chicken eggs and, using appropriate amounts and concentrations of both thyroid and thiourea solutions, compare the resulting growth rate patterns with these chemicals with an untreated control and with the corresponding study of amphibian development.

Students should be encouraged to check out other factors which might exert some influence on embryonic development. Some should design a project in which the effects of an exogenous application of the amino acid tyrosine (precursor of thyroxine) on chick embryos can be investigated and the resultant developmental pattern compared with that of an untreated control.

Few developmental studies elicit as much enthusiasm as do those involving artificial parthenogenesis. Students should investigate and practice meticulously the techniques for obtaining and treating unfertilized frog eggs to induce their development. The resulting embryos with haploid characteristics will make for an interesting comparison with diploid tadpoles produced in the normal manner.

Project—BIOCHEMISTRY OF SOYBEAN SEEDLING DEVELOPMENT

MOTIVATION

An embryo is a progressively changing system engaging in a

continuous pattern of growth. As it develops, increasing its size and complexity, this dynamic system metabolizes available protein-rich nutrients and converts them into embryonic structures. This means that new proteins must be synthesized from the amino acid pool of hydrolyzed protein substances being supplied to the embryo.

Some light may be shed on the biochemical aspects of embryological development through an investigation of the sequence of amino acids appearing in a developing plant or animal. As the amino acids are utilized by the developing organism to synthesize essential nitrogenous organic compounds for its morphogenesis, the chemical changes in the amino acid pool of embryo extract at various developmental time periods can be correlated with the morphological developments occurring during the same period.

MATERIALS REQUIRED

A project study of some biochemical changes involved in the development of soybean seedlings makes use of the technique of paper chromatography and will require, for each participating student group:
> six dried soybean seeds, soaked for one hour;
> six soybean seedlings at each age of germination:
>> one day, three days, five days, eight days;
> a chromatography jar;
> Whatman #1 chromatography paper;
> mortars and pestles;
> five 25 ml. Erlenmeyer flasks;
> five 5 ml. pipettes;
> four seed germinating chambers;
> a chromatographic solvent;
> ninhydrin aerosol spray;
> 85% ethyl alcohol; and
> sodium hypochlorite solutions, 2% and 5%.

A centrifuge should also be available for use.

DEVELOPMENT

Before beginning the project, students should review the basic procedures for paper chromatography and applications of the technique to chemical analyses of amino acids. Also, prior to starting, several preliminary steps should be taken:

1. Germinating chambers should be rinsed with a 2% solution of sodium hypochlorite.
2. Seeds to be germinated should be given a fungicide treatment consisting of soaking in a 5% solution of sodium hypochlorite for 10 minutes before germination.
3. Germination of seeds should be started at appropriate time intervals to provide embryos in seeds soaked for one hour, and seedlings at one, three, five, and eight days germination at the time the project is to be initiated.

Each member of a group can be assigned one embryo age level with which to work and for which to prepare an embryo extract. Seed coats, where still present, should be carefully removed, cotyledons separated, and embryos removed and transferred to the mortar designated for the six specimens representing the given stage of development. Using a small amount of white sand as an abrasive, embryos should be crushed with a grinding action of a pestle. The addition of 5 ml. of 85% ethyl alcohol and continued grinding will yield some embryo extract which can be decanted into a 25 ml. Erlenmeyer flask, after which the addition of another 5 ml. of alcohol and continued grinding of the remaining solids should follow. All extract and solid substances should then be combined in the flask which, properly labeled to indicate the kind and age of the embryo from which its contents have been obtained, should be stored overnight in a refrigerator. After 24 hours embryo extracts should be transferred to separate centrifuge tubes and centrifuged for 10-12 minutes at low-medium speed. The supernatant should then be transferred to clean containers and placed in a hood or other well-ventilated area for evaporation of some of the alcohol by fanning off some of the vapors, until only about 2 ml. of each extract remains. Each embryo extract should then be transferred, via capillary pipette, to its designated position on a sheet of chromatography paper which has been marked, along a line one inch from its base, to receive each of the five extracts for analysis. Using an ethanol-ammonium hydroxide-distilled water (180:10:10) solvent mixture in a vapor saturated chromatography jar, chromatograms should be run, sprayed with ninhydrin, and analyzed. Comparisons of chromatograms produced by the dormant embryos with those of extracts from the different periods of development should be made and, if possible, identification of specific amino acids by comparisons made with a standard.

In a similar manner, some student groups can investigate the biochemistry of lima bean or corn seedling development.

DISCUSSION AND INTERPRETATION

A comparison of the chromatograms produced by each of the developmental stages of a given type of seedling should be made and the significance of the noted differences discussed. By arranging the chromatograms so that they represent a sequence of developmental stages from dormancy to eight-day-old seedlings, students should consider if it is reasonable to expect protein synthesis to be at its maximum in the adult stage of the plant being developed.

The project should be discussed from the standpoint of having used only plant embryos for the analyses; thought should be given to the results to be expected if whole seed extracts had been used. This might well stimulate the formulation of plans for a project to check out the prevailing hypothesis.

The chemistry involved in ninhydrin reactions for locating the amino acids and the importance of a standard of known amino acid ninhydrin patterns for identification of the ninhydrin reacting substances in the analyses should be discussed.

Application of the project techniques to a determination of biochemical changes during the development of animal embryos should also be considered and suggestions made for the use of black flies, fruit flies, or grasshoppers as experimental organisms for such a project study.

FURTHER STUDY

Students should be encouraged to plan projects in which biochemical changes can be correlated with morphological development in insects and crustaceans. Others wishing to apply the study to higher organisms should design and conduct a project which correlates amino acids to the appearance of specific structures such as hind legs and fore limbs in the development of frog tadpoles.

Further investigation into the use of chromatographic solvents should also be encouraged. Students should repeat some of the procedures for paper chromatography in which substitutions of other solvents are used and an evaluation of the various substances in specific situations tabulated for reference.

Thin layer and gas chromatographic techniques should also be investigated and, where possible, demonstrated to the class by those members who have experience in their use.

Project—INHIBITION OF TERATOGENIC ACTIVITY

MOTIVATION

Although the use of certain drugs has long been associated with fetal loss during pregnancy, it was not until the tragic consequences of the thalidomide episode of the 1960's that damage to developing embryos was fully recognized as a possible side effect produced by some medications reputed to be "safe." The use of thalidomide as a mild sedative by mothers-to-be caused their unborn children, in critical stages of development, to react in alarming fashion; when exposed to the chemical influence during periods of extreme vulnerability, leg and arm buds developed abnormally and produced limb deformities in the newborn.

Today the developing embryo is identified as an organism with unique characteristics which distinguish it from the fully developed form of the same species, and drugs are viewed more readily as agents that may be harmful to embryonic development and responsible for the production of congenital defects. Problems concerning how to combat the teratogenic activity of suspect drugs and escape the subsequent damaging effects exerted on future generations challenge the skill and ingenuity of researchers in this field.

MATERIALS REQUIRED

A project which investigates the teratogenic activity of several disazo dyes, alone and in combination, during the embryogenesis of the chick will require:

6½ dozen fertile white leghorn chicken eggs;
fresh 0.1% saline solutions of disazo dyes

Trypan blue,
Niagara blue 2B, and
Congo Red;
fresh 0.1% saline solution;
an egg candler;
an egg incubator;
sterile 1 cc. tuberculin syringes;
sterile #20 gauge hypodermic needles;
cotton;
alcohol;
foreceps;
culture bowls;
egg drills; and
dissecting microscopes.

DEVELOPMENT

There is great advantage in advance planning and preparation
for this project study. Eggs to be used should be candled to make
certain that they are fertile and the temperature (103° F) and
humidity (56%) of the egg incubator regulated and maintained at
these optimum levels for 2-3 days prior to introducing the eggs. If
an entire class is to participate in the project, the formation of six
student groups will allow for each group to inject eggs, in
duplicate, with the full complement of six substances in the series,
or each group to be assigned one of the substances to inject into
each of a dozen eggs. In either case, results from the entire class
should be pooled for analysis and evaluation of the study.

Eggs should be incubated at 103° F and turned twice daily to
prevent adhesion of membranes. During this period students
should use grocery-store eggs and distilled water to practice the
proper technique for making injections through egg shells and
should prepare, dispense into small vials, and sterilize a 0.1% saline
solution and 0.04% solutions in 0.1% saline of disazo dyes Trypan
blue, Niagara blue 2B and Congo red.

At the end of 36 hours of incubation, eggs should be removed
from the incubation zone, candled to identify the positions of
embryo and yolk sac, and lightly marked on the shell with lead
pencil to outline the location of the yolk sac. Properly marked for
identification corresponding to the substance to be injected, eggs

should then be swabbed with alcohol-moistened cotton at the sites
to be injected and their shells penetrated, directly above the yolk
sacs, with a sterile egg drill. Using a sterile 2 cc. tuberculin syringe
fitted with a #20 gauge hypodermic needle, chemical solutions,
previously prepared and assigned to groups, should be injected
into the yolk sacs to yield—

one dozen eggs treated with 0.25 cc. Trypan blue solution,
one dozen eggs treated with 0.25 cc. Niagara blue 2B
solution,
one dozen eggs treated with 0.25 cc. Congo red solution,
one dozen eggs treated with 0.25 cc. Niagara blue 2B solution
followed by 0.25 cc. Trypan blue solution,
one dozen eggs treated with 0.25 cc. Congo red solution
followed by 0.25 cc. Trypan blue solution, and
one dozen eggs treated with 0.25 cc. saline solution, to serve
as a control group.

Six eggs should be left untreated for completely undisturbed
normal embryo development.

After treatment, eggs should be sealed at the injection sites with
melted paraffin wax, returned to the incubator, and the incubation period continued for an additional eight days. Embryos at 11
days of development should then be sacrificed: place an egg in a
nest of cotton with air sac oriented up, crack the shell with sterile
forceps, remove shell fragments and exposed membranes, and
gently pour the shell content into a clean culture bowl. The
embryo and its extra-embryonic membranes should be examined,
after which the embryo should be carefully lifted with a pair of
forceps placed under the head, and transferred to a culture bowl
containing clear saline solution where, with the aid of a dissecting
microscope, a comparison of treated and untreated specimens can
be made. Embryo mortality and all malformations—eye and beak
deformities, the rumpless condition, etc.—should be noted and
recorded on a chart which provides for the tabulation of mortality
rate and kind and frequency of abnormalities observed among the
survivors of embryos exposed to each of the test substances. An
analysis of the data will reveal pertinent information for determining the effect of each dye injection, alone and in combination with
another, on the embryogenesis of the chick and for the use of

non-teratogens as potential inhibitors of the activity of agents producing teratogenic effects.

DISCUSSION AND INTERPRETATION

The project should be considered for its scientific merit as well as for its results. The techniques employed, conditions maintained, adherence to a carefully designed time schedule, and value of running controls with saline injections as well as those totally untreated should be discussed as they contribute to the study.

Results obtained by various student groups should be compared and discussed. Non-active substances as well as those observed to be teratogenic in their activity should be identified and special emphasis focused on combinations in which one substance, observed to be inactive when operating alone, also appears to provide some measure of protection to the embryo against the teratogenic action of another, as expressed by the effects produced when it is the sole agent. The value of these substances as potential inhibitors of teratogenic action brought about by chemical agents that come in contact with an organism during its embryogenesis should be carefully assessed and their modes of action studied for this and other specific instances. In this relation, research studies involving small mammals in which subcutaneous injections of disazo dyes into pregnant females indicated interference of an inactive dye with one of known teratogenicity should be investigated and reported to the class.

The suitability of embryos for injection in their 36th hour of development should be evaluated in terms of the project findings. Students should be encouraged to suggest steps that might be taken to verify this as the most critical time in which teratogenic effects are exerted on the chemical differentiation of the tissues witnessed to be affected by the agents tested and, via a complete study of normal embryogenesis, to pinpoint periods of vulnerability of other structures.

Applications of studies involving damage during embryonic development should be considered as they relate to human experience; a high incidence of fetal loss among untreated diabetic pregnant women, congenital sight and hearing defects in children born to women suffering German measles during early months of pregnancy, and the congenital deformities due to thalidomide usage and cortisone intoxication of prospective mothers point up the role of the unborn baby's environment in influencing its development. The implications for untested drugs—LSD for one—and the immediacy of the need for identifying chemical agents which might act to interfere with the teratogenic effects of chemicals known to cause them should be emphasized.

In light of the multitude of medical preparations, drugs, and chemicals available to the public today and of the extreme vulnerability of the embryo, the viewpoint so aptly expressed by Dr. Frances Oldham Kelsey that "the fetus or newborn may be, pharmacologically, an entirely different organism from the adult" is worthy of discussion and due consideration.

FURTHER STUDY

A more comprehensive study of the teratogenic effects of related disazo dyes can be made by incorporating Evans blue, Azo blue and Niagara blue 4B into the basic study series. In addition to the wider spectrum of dyes provided, if all are used, greater flexibility will be gained in the selection of new groupings and combinations of test agents or for making substitutions, if desired.

A variation of the basic project can be designed to test the same group or other groups of dyes by directing the injections into the subgerminal cavity and comparing the results with those obtained when injections were made into the yolk sac of the egg. Pesticides also can be tested as teratogenic agents.

Students wishing to pinpoint the time during which a chick embryo is most vulnerable to a teratogenic agent should make an in-depth study of a single disazo dye, such as Trypan blue, in which fertile eggs are injected into the yolk sac at various time periods—12, 24, 36, 48, 72, and 96 hours of incubation—to determine the mortality rate and kind and frequency of malformed embryos caused by the injections at these points of development. For comparison, another group can plan a similar project, making injections into the subgerminal cavity in an otherwise identical procedure. Results of the two projects, run in parallel, should be graphed on the same axis for analysis and interpretation.

Students who have had some experience in handling mice or rats might extend the project study to include small mammals.

They should be encouraged to design and conduct a project in which injections of Trypan blue solutions are made intraperitoneally into pregnant specimens and the sacrificed embryos are examined for deformities. This project can be expanded to include an investigation into the possible interference of one disazo dye with the activity of another, thus checking their uniformity of pattern whether their effects be exerted on chick or mammalian embryos.

Highly motivated students should engage in more individualized projects such as an investigation into a possible correlation between the amount of a teratogen and the severity of the malformations it causes in chicks or other experimental embryos. Others should design a project in which sedatives, cortisone, and other drug preparations are tested, via injections into small mammalian experimental organisms, for their teratogenic activity. For the very bright student, an original project study which seeks to discover an inhibitor for a known teratogen will provide a wholesome challenge.

RECOMMENDED READING

Davidson, Eric H. "Hormones and Genes," *Scientific American,* (1965), 212, 6.

Ebert, J.D. *Interacting Systems in Development,* Modern Biology Series, Holt, Rinehart and Winston, New York, 1965.

Edwards, R.G. "Mammalian Eggs in the Laboratory," *Scientific American,* (1966) 215, 2.

Eigsti, Nicholas W. "Paper Chromatography for Student Research," *The American Biology Teacher,* (1967), 29, 2.

Etkin, William. "How a Tadpole Becomes a Frog," *Scientific American,* (1966), 214, 5.

Fischberg, M. and A. W. Blacker. "How Cells Specialize," *Scientific American,* (1961), 205, 3.

Frieden, Earl. "The Chemistry of Amphibian Metamorphosis," *Scientific American,* (1963), 209, 5.

Grabowski, Casimer T. "Lactic Acid Accumulation as a Cause of Hypoxia-Induced Malformations in the Chick Embryo," *Science,* (1961), 134, 3487.

Gray, George W. "The Organizer," *Scientific American,* (1957), 197, 5.

Gurdon, J.B. "Transplanted Nuclei and Cell Differentiation," *Scientific American,* (1968), 219, 6.

Hadorn, Ernst. "Transdetermination in Cells," *Scientific American*, (1968), 219, 5.

Hais, I.M. and Macek. *Paper Chromatography*. Academic Press, New York, 1963.

Hamburger, V. *A Manual of Experimental Embryology*. University of Chicago Press, Chicago, 1947.

Koller, Dov. "Germination," *Scientific American*, (1959), 200, 4.

Konogsberg, Irwin R. "The Embryological Origin of Muscle," *Scientific American*, (1964), 211, 2.

Mazia, Daniel. "How Cells Divide," *Scientific American*, (1961), 205, 3.

Moog, Florence. *Animal Growth and Development*. BSCS Lab Block, D.C. Heath and Company, Boston, 1963.

———— "Up From the Embryo," *Scientific American*, (1950), 182, 2.

Needham, J. *Biochemistry and Morphogenesis*. Cambridge Press, New York, 1942.

Perry, R.P., R.R. Srinivasan, and D. E. Kelley. "Congenital Anomalies in Hamster Embryos with H-1 Virus," *Science*, (1964), 145, 3631.

Roth, J.S., G. Buccino and N.W. Klein. "Inhibition of Growth of Chick Embryo by Inhibition of Deoxycytidylate Deaminase," *Science*, (1963), 142, 3598.

Rugh, Roberts. *Experimental Embryology*. Burgess Publishing Company, Minneapolis, 1962.

Stein W.H. and S. Moore. "Chromatography," *Scientific American*, (1951), 184, 3.

Sussman, M. *Animal Growth and Development*. Prentice-Hall, Inc., Englewood Cliffs, N.J., 1960.

Taussig, Helen B. "The Thalidomide Syndrome," *Scientific American*, (1962), 207, 2.

Waddington, C.H. "How Do Cells Differentiate?" *Scientific American*, (1953), 189, 3.

Wigglesworth, V.B. "Metamorphosis and Differentiation," *Scientific American*, (1959), 200, 2.

Williams, Carroll M. "The Metamorphosis of Insects," *Scientific American*, (1950), 182, 4.

———— "The Juvenile Hormone," *Scientific American*, (1958), 198, 2.

Witschi, E. *Development of the Vertebrates*. W.B. Saunders, Philadelphia, 1956.

5

NUTRITION

Nutrition, in a broad sense, is concerned with providing for the energy needs of an organism while supplying its requirements for building and repairing tissues and regulating its life processes. While all rely upon the environment to supply some of the materials for these purposes, organisms vary significantly in the degree to which they express their dependence. The importance of adequate nutrients, however, has long been recognized; the problem of malnutrition has plagued man since the dawn of history and the far-reaching effects of deficiencies which limit plant growth and vigor are all too familiar to man the world over. Recently, the nutritional aspects of atherosclerosis, diabetes, and other human disorders have presented new insights into the role played by dietary excesses as well. Deleterious effects of high levels of proteins, fats, carbohydrates, vitamins, and minerals are constantly being revealed, and nutritional studies which investigate mechanisms for avoiding and/or counteracting these effects are being rapidly accelerated in this dynamic area of biological research.

Project—GROWTH PROMOTING NUTRIENTS IN BACTERIA

MOTIVATION

The presence of adequate and proper nutritional materials is generally reflected in the phenomenon of growth; nutrients are essential for providing both the substance and structure of all living systems.

The absence of suitable building materials during critical periods of an organism's lifetime can result in tragedy and disaster. In the case of the nutritional deficiency disease known as *Kwashiorkor*, prevalent in the under-developed countries of the world today, small children afflicted with the disease fail to grow properly— their bellies become swollen and filled with fluids, limbs become shrunken, muscles become atrophied, and minds become dull and listless. If they are not treated they will die. Treatment for this disease is an amazingly simple application of principles of good nutrition; daily supplements of as little as six ounces of dried skim milk or other quality protein will bring about a remarkable recovery and, in most cases, properly treated children will grow into normal healthy adults.

MATERIALS REQUIRED

The study of nutrition can be approached from the direction of nutritional requirements for bacterial growth and will require for each student or group engaging in the project:

one flask of each of four differential nutrient media: saline, glucose, peptone, and vitamin; 50 ml. each;
one flask containing 50 ml. sterile distilled water;
one tube containing 10 ml. sterile distilled water;
one 24 hour agar slant culture of *Escherichia coli;*
one inoculating loop;
one sterile 1 ml. pipette; and
ten sterile petri dishes.

The use of a 37° C incubator, Quebec colony counter, 45° C water bath, refrigerator, and an autoclave or pressure cooker will also be required for use in the project study.

DEVELOPMENT

As a preliminary procedure, each student group should be assigned the preparation of a different media, to be shared with other groups engaging in the investigation. Four media should be prepared:

Saline media: 9 ml. NaCl in 1000 ml. distilled water plus 15
gm. agar;

Glucose media: 25 gm. glucose in 1000 ml. distilled water
plus 15 gm. agar;

Peptone media: 25 gm. peptone in 1000 ml. distilled water
plus 15 gm. agar; and

Vitamin media: 6 multi-vitamin pills dissolved in 1000 ml.
distilled water plus 15 gm. agar.

In each case, ingredients should be heated gently to boiling,
allowed to boil for one minute, dispensed in 50 ml. quantities in
150 ml. Erlenmeyer flasks, and sterilized by autoclaving. Until
needed, all media, properly labeled, should be stored in a refrigerator. In a similar manner, distilled water should be dispensed in 50
ml. and 10 ml. quantities in flasks and tubes, respectively,
sterilized, and stored for future use.

On the day preceding the beginning of the project study,
transfers of *Escherichia coli* from a stock culture to agar slants,
one per student group, should be made and incubated at 37° C for
20-24 hours.

When the project is to be started, all flasks containing media
should be placed in a water bath and, while the media is being
adjusted to 45° C, the inoculum should be prepared. Using sterile
technique, students should transfer one loopful of the 24 hour
slant culture of the test organism, *Escherichia coli,* to a sterile
distilled water blank and, with gentle agitation, distribute the
organisms to make a uniform suspension. Again employing sterile
methods, an inoculum of 0.5 ml. of the suspension should be
transferred to each of five sterile petri dishes, using a sterile 1 ml.
pipette for the purpose. Plates should be labeled for each of the
media to be tested and distilled water which will be placed in the
fifth in the series. An additional five plates, receiving no organisms, should be properly labeled to serve as controls for each of
the test media plates.

45° C media should be poured into appropriately marked plates
(one test and one control for each media) and swirled gently in
the test plates to distribute the organisms of the inoculum
uniformly. After all media has solidified, plates, properly labeled
for identification, should be taped and incubated, in an inverted
position, at 37° C for 24 hours.

After the incubation period, plates should be examined for growth, using a Quebec colony counter, if necessary, to determine the number of colonies produced in each media. If immediate counting is not convenient, plates may be refrigerated until conditions are more suitable. Results should be tabulated, for each media, in chart form for an analysis and evaluation.

DISCUSSION AND INTERPRETATION

Each group should determine the comparative value of the substances tested as growth promoting nutrients for *Escherichia coli* and the findings of all groups should be examined for evidence of uniformity and agreement. The role of protein as a growth food should be stressed and the expected results of a proposed project including some additional media such as glucose-peptone, 3:2; glucose-peptone, 4:1; and glucose-peptone-vitamins, 3:1.5:0.5 predicted.

The use of "trash fish" flour, containing 80% protein, should be explored and discussed as a food source and practical solution for feeding the protein-starved peoples of the world. Students may have read of some nutritional studies that have been conducted using fish flour and should be encouraged to report the findings of these studies to the class.

The experiment of the late 1940's, conducted on the Bataan peninsula, should also be researched and reported to the class for a discussion of the nutritional values of white rice enriched with a chemically prepared thiamin-niacin-iron compound as compared with the control diet of polished white rice considered "normal" in the area at the time. In this respect, the nutritionally incomplete nature of the three major cereals—rice, wheat, and corn—each of which lacks one or more of the essential amino acids, should be discussed and, in view of the successful results reported from the practice of fortifying animal feeds, some recommendations made for fortifying these grains for human consumption. Personal diets should also be scrutinized and evaluated in terms of bodily requirements during childhood and adolescence. A diet of candy bars and soda pop should come under the closest scrutiny. Perhaps a debate on the pros and cons of nutritive values of Tiger's milk (really a nutrition booster containing yeast and soy beans) and of other exotic health fad foods can be planned after researching the substances and basing arguments on scientifically based research studies.

The advantages of using microorganisms as the experimental organisms in the project on nutrition should be assessed, and disadvantages should be cited as well and weighed against the advantages stated.

The technique of the pour plate-count method as a measure of bacterial growth should be compared with that of visual turbidity in liquid medium and, if a spectrophotometer is available, the amount of light transmission through broth medium as an indicator of the density of organismal growth should be demonstrated.

FURTHER STUDY

The possibilities for study in the area of nutrition are almost without limit. The basic project can be repeated using other microorganisms as experimental subjects; a spectrum of proteins—gelatin, albumin, and casein—can be investigated in a qualitative study of different proteins and their growth promoting abilities; specific amino acids, alone or in combination, can be tested for a determination of nutritional requirements for a given microorganism's growth; and, by extending the basic project to include nutritional substances in combination, the nutrients which provide optimal growth potential for an organism can be determined. Individually or in groups, students should pursue these studies, using bacteria, yeasts, and molds as the experimental organisms.

Some students may be assigned alternate techniques by which to measure organismal growth. The visual turbidity of liquid media technique should be researched thoroughly and, run in parallel with the pour plate-count technique, compared and used as a confirmation test of the results obtained in the basic study.

The nutritional value of aquatic plant life should be investigated by some individuals. They should be encouraged to design and conduct a project in which duckweed is used as the sole food source for a group of laboratory mice whose growth should then be compared with that of a control group fed a conventional well-balanced diet.

Advanced students should investigate and employ auxano-graphic techniques and, utilizing combinations of various classes of nutrients, determine the nutritional requirements of the mold *Phycomyces* or other experimental organism.

Current nutritional problems of man offer many challenges to students. Those who are highly motivated should design projects which investigate the role of high animal product diets in the production of high cholesterol levels and atheromatous plaques and the high levels of calcium and vitamin D in the diet which lead to calcified tissues and conditions associated with hypervitaminosis, using white mice as the experimental organisms for the study. For an additional activity, students should plan a project in which fruit flies are grown on carbohyrate and water media with and without yeast to determine the role played by protein in the reproductive process, as exhibited by egg production in the cultures.

Project—MINERAL DEFICIENCY IN GROWTH OF DWARF TOMATO PLANTS

MOTIVATION

Although plants and animals differ markedly in the manner in which they secure minerals, both depend upon certain inorganic elements to satisfy their requirements for life. Neither can grow or function normally in the absence of any of the essential minerals, and malfunctions due to their absence or deficiency can be corrected only by their addition in appropriate kind and quantity.

Plants grown hydroponically in nutrient solutions can be used to identify the mineral elements essential for their normal growth and appearance. So readily recognizable are the characteristic deficiency symptoms of plants grown in solutions known to be lacking in specific chemical content that the growth associated with nutrients in solution can be used as an index of both macro and micro mineral elements. Diagnoses for specific unhealthy appearance and growth patterns due to mineral-deficient diets of rats, rabbits, and chickens also make possible the use of small animals for qualitative bioassay. Biologists thus make wide application of the techniques for investigating problems relating to mineral and other nutritional deficiencies in plants, animals, and humans, while seeking to discover ways to improve their health and vigor.

MATERIALS REQUIRED

Mineral requirements of plants can be determined by hydroponics techniques used for the growth of seedlings. A project for this determination will require:

ten tomato seedlings, Burpee's Tiny Tim or other dwarf variety;

five one-gallon wide mouth jars;

five jar covers to fit, with ½ inch diameter holes;

five bulb type aerators;

aluminum foil;

foam rubber or glass wool;

a fluorescent light source;

distilled water;

mineral stock solutions:

$MgSO_4$	- 7.8 gm./1000 ml. distilled water
$CaH_2(PO_4)_2$	- 4.0 gm./1000 ml. distilled water
KNO_3	- 7.8 gm./1000 ml. distilled water
$CaSO_4$	- 12.2 gm./1000 ml. distilled water

mineral trace element stock solutions:

#1	$MnSO_4$	- 1.0 gm. in 250 ml. warm distilled water
	$ZnSO_4$	- 1.0 gm. in 250 ml. warm distilled water
	$CuSO_4$	- 1.0 gm. in 250 ml. warm distilled water combined to form a stock solution
#2	$FeCl_3$	- 1.0 gm. dissolved in 500 ml. warm distilled water

DEVELOPMENT

While seeds selected for growth are being germinated in moist vermiculite, mineral nutrient solutions should be prepared, as indicated, by each student group.

Group 1, assigned a mineral-sufficient nutrient solution, should combine in a one gallon jar:

600 ml. of each stock solution: $MgSO_4$

$CaH_2(PO_4)_2$

$$KNO_3$$
$$CaSO_4$$

1 ml. stock mineral trace solution #1 and
5 ml. stock mineral trace solution #2 with
sufficient distilled water to make one gallon
of a mineral-sufficient nutrient solution.

Other groups should use variations of the same basic formula, making alternations, where necessary, to prepare the designated mineral-deficient nutrient solutions.

Group 2, assigned a nitrogen-deficient solution, should substitute KCl (3.5 gm./600 ml. distilled water) for the KNO_3 solution;

Group 3, assigned a phosphorus-deficient solution, should substitute $CaCl_2$ (3.5 gm./600 ml. distilled water) for the CaH_2 $(PO_4)_2$ solution;

Group 4, assigned a potassium-deficient solution, should substitute $NaNO_3$ (4.0 gm./600 ml. distilled water) for the KNO_3 solution; and

Group 5, the control group, should use distilled water only.

All nutrient solutions should be adjusted to pH 5.7 in correspondingly labeled culture jars equipped with bulb type aerators and covered around the outside with aluminum foil. Fluorescent lighting should then be supplied equally to the five culture jars in the series.

When the seedlings have grown to a length of 3-4 inches, care should be taken to select plants of approximately the same size and appearance for transfer to the growth chambers containing differential nutrient solutions. After the stems have been placed through the holes in the jar tops, small pieces of glass wool or foam rubber, strategically placed at the sites, will support the seedlings and secure their positions which allow the root systems to come in contact with the nutrient solutions.

In each culture jar, the aerator should be pumped about 18-20 times to bring air in contact with the roots of the seedlings. This practice should be repeated twice daily, morning and afternoon, until the roots become long enough to allow for an air space above the root level in the jar. The water level in the jar should be checked daily and water lost, due to evaporation or plant use, replenished by the addition of distilled water. Every 8-10 days, the entire root system should be removed from the solution and gently rinsed with distilled water to remove all traces of chemical deposit build-up or other debris. The nutrient solution should then be replaced with a fresh supply of the same chemical composition and the plant returned to its position for continued growth. This process should be repeated every 8-10 days for a period of 3-4 months, during which time the plants should be examined regularly. Measurements and notes concerning the general appearance should be recorded on a chart which provides for entries for plants grown in all nutrient solutions included in the project study.

DISCUSSION AND INTERPRETATION

The tomato plants grown in the various mineral-deficient solutions should be compared with those grown in the full nutrient mineral-sufficient solution. Based on the size and appearance of each plant, the role of each nutrient should be established as it contributes to the nutritional requirements of tomato plants.

The reduced value of mineral-deficient plants as sources of animal and human food should also be considered, with emphasis placed on the importance of the green plant as a basic source of food for all heterotrophs.

Techniques employed, such as the use of aluminum foil to cover the culture jars, the growth of some plants in distilled water, the use of fluorescent lighting, and the routine aeration of plant roots should be discussed as they contribute to the project.

The importance of minerals in animal nutrition also should be evaluated, and the controversy over the introduction of fluoride to drinking water supplies researched and debated in class.

FURTHER STUDY

Using lima bean, corn, marigold, or other seedlings, the project can be repeated or run in parallel with the basic study for a confirmation of the basic mineral requirements and/or determination of specific needs of certain plant forms.

Some students may wish to explore the effects of mineral deficiencies in animal diets. They should investigate the use of mineral-deficient and mineral-sufficient rat diets and, using these prescribed substances, compare the growth and appearance of rats to which they have been fed. By keeping detailed records of their observations, students engaging in such a project will be able to identify the symptoms associated with the specific mineral deficiencies and then, by supplementing the diet with minerals that were lacking, to restore the animals to a state of good health.

The importance of metals to nutrition also elicits much student enthusiasm. Highly motivated students should be encouraged to plan a project in which the effects of varied diets—deficient in zinc, copper, iodine, sulfur, and manganese—are studied, using chickens as the experimental subjects.

More advanced students should be encouraged to investigate the effects of mineral deficiencies on the growth of *Lemna perpusilla.* After having researched the mineral composition of recommended growth media, they should design and conduct a project which investigates growth patterns of the organism in media deficient in potassium, phosphorus, and magnesium, comparing it with the normal pattern resulting from growth in media containing complete mineral requirements for its optimum growth.

Project—FOLIAR ABSORPTION IN PLANT NUTRITION

MOTIVATION

Automated sprinkling systems, when employed as the sole mechanism for supplying water and mineral fertilizers to plant crops grown in large commercial greenhouses, produce highly satisfactory results. Application of the mineral nutrients to plants in a fine mist creates an atmosphere mildly suggestive of conditions found in western Java where the annual rainfall of about 72 inches and mean temperature of 78° F maintain a constant high humidity and lush plant growth that is essentially continuous. It appears that, in addition to that portion of the solution which comes to rest on the soil for normal root absorption, the mineral rich mist enveloping the plants is also absorbed by their above-ground-level structures. Of primary importance to the commercial plant grower, this principle can be adapted to other situations where an accelerated growth of healthy, well-nourished plants is desired.

MATERIALS REQUIRED

A project which investigates foliar absorption by plants and compares the resulting growth with that of plants engaging in absorption by root systems only will require:

    large size planters with plastic covers;
    bean, pea, radish or other seedlings;
    a botanical mist sprayer;
    vermiculite or sphagnum moss;
    plant fertilizer, 18-18-18 or 20-20-20; and
    distilled water.

DEVELOPMENT

The project lends itself to a class study in which both experimental and control groups work cooperatively as a team to study the growth of plants with and without the employment of foliar absorption of water and minerals in solution. Each team, consisting of two groups of students, should select seedlings of approximately the same size and appearance within a given plant type and, dividing them equally, transplant them in moist vermiculite in two identical planters with plastic covers.

A stock mineral solution, prepared by dissolving 12 gm. of fertilizer of either formula 18-18-18 or 20-20-20 in one gallon of distilled water, should be maintained throughout the run of the project for use on both the experimental and the control groups of seedlings.

Using a botanical mist sprayer, the experimental group should apply, each morning, one liter of the fertilizer solution to the experimental seedlings. If possible, this should be accomplished without disturbance to the plastic cover.

Similarly, but by introducing the fertilizer solution to the vermiculite only, the control group also should apply one liter of the same solution at the same time to the control plants. Again, the plastic cover should not be unduly disturbed.

This procedure should be continued daily for a 2-3 month period, during which time records should be kept concerning the size, appearance, growth rate, and mortality rate for both the experimental and the control plants.

DISCUSSION AND INTERPRETATION

Discussion of the results should compare the experimental with the control group for each plant type. A comparison of the growth rates should be used as a measure of the absorption of nutrients and, in summary, the relative merits of foliar absorption and the more conventional method of root system absorption only should be established.

Foliar absorption in carnivorous plants should be considered also. If a pitcher plant is available its leaves should be examined for the presence of liquid and insect remnants (in some stage of degradation) for absorption into the plant tissues. Experiments which showed that ten nitrogenous compounds were absorbed more rapidly than water should be researched and discussed as a phenomenon which gives advantage to such plants by allowing them to absorb some amino acids ready-made. Having thus determined the value of foliar feeding to the insectivorous plants, insects in the diet should be equated with the growth of young plants and the production of flowers in mature plants of these species.

FURTHER STUDY

Students who are intrigued by plants that practice carnivorousness should make a study of how this characteristic contributes to their nutrition. A comparative study of the growth and appearance of pitcher plants deprived of insects in the diet and those fed some carpenter ants should be undertaken to determine the role of these mineral rich foods obtained by the plant via foliar feeding.

Students should be encouraged to collect and bring in samples of soil in which pitcher plants, venus flytraps, and sundew plants have been found growing in nature. After investigating the proper techniques for the analyses, they should analyze the mineral content of this soil and compare it with an analysis of garden soil. Such a project seeks to discover why carnivorous plants must be equipped to obtain mineral substances by processes other than root absorption.

The area of leachate nutrition also should be explored. Students should design and conduct a project in which a nutrient solution extracted from pitcher plant leaves, leached for 24 hours in distilled water, is evaluated by the resultant growth of pitcher plants watered with this solution, using a similar plant watered with a mineral-sufficient solution as a control.

Students should also be encouraged to investigate other techniques which might be employed to illustrate foliar absorption. Some of the more advanced students might plan a project in which radio-isotopes of zinc, calcium, iron, and iodine are fed to carpenter ants which, in turn, are fed to pitcher plants.

Project—GALACTOSEMIA IN WHITE RATS

MOTIVATION

Although all heterotrophic organisms, including man, are dependent upon preformed organic molecules for their nutritional needs, the form in which foodstuffs are generally available is not suitable for immediate use by the cell's machinery. Organisms must be equipped with a mechanism for processing these materials to meet the specifications of their cellular requirements. The process involves both physical and chemical changes and is progressive, involving many enzymes which are highly specific in their action and including the formation of many intermediate products. As with any system, functional defects occurring at any step may block the formation of the final product. Often intermediate products of suitable molecular size pass across membranes but, being unusable, accumulate and cause abnormal conditions which can manifest themselves in the form of growth retardation, mental deficiency, jaundice, digestive disorders, and eye cataracts. Such functional defects, if identified early, can be compensated for by dietary controls which bar or severly limit the intake of foods containing substances with which the body is unable to cope.

MATERIALS REQUIRED

A project which investigates how galactosemia can be induced and subsequently treated will require:

ten or 12 weanling rats from the same litter;
ten or 12 animal cages;
ten or 12 animal water bottles;
ten or 12 animal feeding cups;
an animal weighing scale;
a 70% galactose rat experimental diet; and
a galactose-free rat control diet.

DEVELOPMENT

While rats are approaching the proper age for participation in the project study, experimental and control diets should be prepared according to the formulas:

Experimental		Control
20%	Casein	20%
70%	Galactose	
	Glucose	70%
8%	Peanut Oil	8%
2%	Minerals	2%

Both diets should also include multi-vitamin supplements.

When three weeks old, rats from the same litter should be weighed and placed in individual cages. Half of the specimens, designated as the experimental group, should be cared for by one group of students who should feed them ample daily rations of the experimental diet, provide fresh water twice daily, and clean the cages every second day. Another group of students, caring for the control specimens, should handle their charges with the same good care, only feeding them the control diet. Both groups, working as a team, should note and record for the specimens being studied the amount of water intake, weight gain or loss, presence or absence of galactose in the urine, appetite, behavior pattern, and appearance of the eyes, fur, and body carriage. These data, collected daily over a six week period, should be recorded on a composite chart for a comparative study.

Specimens showing marked physical impairment should be removed from the experimental diet and the study continued, with records also continuing after their change to a galactose-free diet.

DISCUSSION AND INTERPRETATION

Evidence of an induced galactosemic condition in members of the experimental group should be discussed and related to the natural occurrence of the disorder in which, due to an inborn error, individuals (including humans) are unable to metabolize galactose. This phenomenon should be researched fully and importance, placed upon its early diagnosis and treatment, for, while the elimination of milk (the source of lactose from which galactose is derived) from the diet of infants will bring about a regression of most of the symptoms, mental retardation, once established, makes a permanent impression.

The chemical nature of galactose and glucose also should be noted. Having the same chemical formula but different molecular configuration, these two carbohydrates differ in that galactose, unless converted enzymatically into glucose, cannot be metabolized by cells. Unused, it accumulates to produce the functional defect galactosemia. The hereditary aspects of the disorder naturally due to a deficiency in the amount of uridyl transferase normally present in the liver, should be investigated.

Other genetically based disorders with nutritional overtones should be researched and reported to the class. Phenylketonuria, in which there are high levels of phenylalanine in the blood stream due to the inability of the body to convert phenylalanine to tyrosine, should be discussed as well as special phenylalanine-free diets such as the commercially prepared Ketonil for its treatment.

FURTHER STUDY

Students wishing to vary the procedure for restoring to good health those rats in which galactosemic symptoms have been produced may treat the afflicted individuals with some of the commercially available milk substitutes. By feeding the diseased rats a diet of Nutramigen or a soybean milk preparation, they can compare the effectiveness of these products in the treatment and cure of induced galactosemia.

Advanced students should be encouraged to plan a project which investigates the nutritional capabilities of microorganisms. Employing microbiological techniques, they should determine microbial utilization of nutrients such as lactose, dextrose, malonate, and citrate, using *Serratia marcescens*, *Pseudomonas fluorescens*, *Aerobacter aerogenes*, *Proteus vulgaris*, *Escherichia coli*, *Bacillus subtilis*, *Sarcina lutea* and *Staphylococcus aureus*, or other spectrum of bacteria as the experimental organisms.

RECOMMENDED READING

Asimov, Isaac. *The Chemicals of Life*. New American Library of World Literature, New York, 1962

NUTRITION

Bogert, L. J. *Nutritional and Physical Fitness.* W.B. Saunders, Philadel
1960.

Brozek, Josef. "Body Composition," *Science,* (1961), 134, 3483.

Crampton, E.W. and L.E. Lloyd. *Fundamentals of Nutrition.* W.H. Freeman
Company, San Francisco, 1960.

Free, Montague. *Plant Propagation in Pictures.* American Garden Guild and
Doubleday & Co., Garden City, New York, 1957.

Galston, A.W. *The Green Plant.* Prentice-Hall, Inc.. Englewood Cliffs, New
Jersey, 1968.

Hammond, Winifred. *Plants, Foods, and People.* The John C. Day Co., New
York, 1964.

Kramer, P. J. *Plant and Soil Water Relationships.* McGraw-Hill Book Company, New York, 1949.

Leaf, Albert L. "Pure Maple Syrup: Nutritive Value," *Science,* (1964), 143,
3609.

Miller, Phillip E. "Zinc Environment Effects on Tobacco for Laboratory
Demonstration," *The American Biology Teacher,* (1968), 30, 7.

Pollack, S., R.M. Kaufman, and W.H. Crosby. "Iron Absorption: The Effect
of an Iron-Deficient Diet," *Science,* (1964), 144, 3621.

Pratt, Christopher J. "Chemical Fertilizers," *Scientific American,* (1965),
212,6.

Rhodes, Lee W. "The Duckweeds: Their Use in the High School Laboratory,"
The American Biology Teacher, (1968), 30, 7.

Schmidt-Nielsen, Knut. *Animal Physiology.* Prentice-Hall, Inc., Englewood
Cliffs, New Jersey, 1964.

Segal, S., H. Roth, and D. Bertoli. "Galactose Metabolism by Rat Liver
Tissue: Influence of Age," *Science,* (1963), 142, 3597.

Syrocki, B. J. "Hydroponics II: A Sand Culture Technique for the Classroom," *The American Biology Teacher,* (1966), 28, 4.

6

METABOLISM

A living organism is a dynamic system. It is constantly engaged in a variety of work processes which are essential for its assimilation, maintenance, growth and reproduction, and must depend upon the outside environment to supply both the energy-yielding and the building-block materials necessary for its continued existence.

Through a mechanism that involves atom or electron transfer from one substance to another, a release of energy may accompany processes of chemical degradation while an energy expenditure is required for activities of chemical syntheses in the living cells. Responding to these chemical reactions occurring within its cell machinery, an organism effects a highly complex system of carefully regulated chemical and physical changes which occur simultaneously and in complete harmony. This is the essence of metabolism.

Project—RESPIRATORY INHIBITORS VS. METABOLISM

MOTIVATION

All living substance is basically alike; it is composed of the same basic elements, utilizes similar organic compounds, and requires

120

energy for engaging in its various life processes. In most of its cells the energy can be derived from glucose, initially by the processes of phosphorylation and glycolysis, and subsequently by a series of enzymatic reactions in which H_2O and CO_2 are formed and energy in usable form is released. The entire process, known as respiration, includes over 20 enzymatic reactions with some energy being released at each step. It involves the passage of hydrogen atoms from one enzymatic substance to another until, finally, when passed to cytochrome oxidase, they combine with atoms of oxygen to form water. It is obvious that any interference with the normal functioning of cytochrome oxidase will prevent the chemical union of hydrogen with oxygen and severely limit the amount of energy derived from food stores. Hence, the rate at which oxygen is taken up by a living system may be viewed as an indication of the amount of energy being derived for metabolic activities, and any agent which acts to lower it appreciably as a respiratory inhibitor.

MATERIALS REQUIRED

An investigation into the effects of several specific chemical substances on the rate of oxygen uptake by a living system can be conducted with easily obtained materials. The following will be needed by each student or group engaging in the project study:

Two wide mouth 500 ml. Erlenmeyer flasks;
glass wool;
plastic or glass tubing;
a mm. scale;
two pinch or screw clamps;
colored ink or food coloring;
15% KOH or NaOH solution with phenolphthalein indicator;
a thermometer;
two 2-hole rubber stoppers to fit Erlenmeyer flasks;
accordion folded filter paper;
a small vial;
a stopwatch or other timing device;
an 18 hour nutrient broth culture of *Escherichia coli;*

1 ml. of formaldehyde; and
other assorted chemical substances.

DEVELOPMENT

A manometer mounted against a mm. ruled scale and connected by way of plastic tubing to flasks at either side should be constructed to serve as a convenient respirometer for use in the project. Both flasks, one designated as a respiratory chamber and the other a temperature control chamber, should be fitted as well with short lengths of glass tubing, equipped with 1½ inch lengths of plastic or rubber tubing extensions that can be closed by tightening screw clamps near their extremities.

After each group of students has assembled a respirometer, they should treat a 25 ml. sample of an 18 hour nutrient broth culture of *Escherichia coli* with 1 ml. of formaldehyde and carefully transfer it to the respiratory chamber which contains a wad of glass wool, previously introduced to the flask via careful handling with long forceps. By means of a bent wire holder attached at its upper end to the rubber stopper of the respiratory chamber, a vial containing 5 ml. of 15% KOH or NaOH solution with phenolphthalein indicator should then be suspended inside the flask at a level of about two inches above the base.

Both flasks should be stoppered tightly and, after equilibrium at room temperature has been reached, screw clamps closed to make the chambers air tight. An initial manometer reading should be taken and recorded on a data chart, with subsequent readings taken every 15 minutes for a period of one hour similarly recorded. All data for the experimental situation and a control, without the formaldehyde treatment, run in parallel, should be graphed in contrasting colors on the same axis for a more comparative analysis.

For a more extensive study, the effects of other chemicals such as sodium azide, carbon tetrachloride, pyrogallol, and mercurochrome or other substances such as carbon monoxide and cyanide can be tested in a similar manner, and, for variation, *Staphylococcus aureus, Bacillus cereus,* or other microorganism can be used as the experimental organism.

DISCUSSION AND INTERPRETATION

The project data should be analyzed and discussed as it relates to the practice of using oxygen uptake as an index of the rate of respiration and as it relates to the mode of action of the chemical substances which act to inhibit an organism's respiration, consequently interfering with its metabolic pathways. The step-wise release of energy from organic molecules and the biochemical aspects of the process of oxidation should be researched and discussed in depth. Students should report their findings concerning the nature of both aerobic and anaerobic forms of respiration and the specific enzyme actions involved in each.

Students should be encouraged to seek information concerning the biochemical activity of dehydrogenase and cytochrome oxidase and to plan and conduct demonstrations involving them before the class. An advanced student can perform a dramatic but simple demonstration in which the rapidity with which the organisms in a 5 ml. sample of an 18 hour broth culture of *Escherichia coli* decolorize 1 ml. of a 1:20,000 solution of methylene blue dye is taken as an indication of the rate at which oxidation, hence cell respiration and metabolism of the organisms, occurs. (Other ratios—4:2, 3:3, and 2:4—may also be used, and in each case, different chemicals added which may act to disable the cytochrome oxidase systems within the living cells.) Demonstrations performed by students should be discussed by the class as extensions of open-ended investigations, and suggestions for still other variations and applications of the studies should be made for individuals or special groups to pursue.

FURTHER STUDY

Multicellular organisms also can be used effectively in similarly designed studies. Projects involving snails, grasshoppers, frogs, sprouting bean seeds, white mice, and hamsters should be planned to determine the effects of temperature, air pressure, magnetic fields, and chemicals on the respiratory rates and to assess the influence of these factors on the rate of metabolism.

The nature of aerobic and anaerobic respiration should be explored, and some of the more advanced students should be encouraged to plan and conduct a project in which the respiratory rates of yeast cells, grown separately under both aerobic and anaerobic conditions, are compared and the metabolic pathways of the products resulting from each mode of cell respiration are determined.

Bright students should be motivated to investigate the succinoxidase system. After researching this intermediate link between oxygen metabolism and the oxygen transport system, they should design a project which investigates the effect of oxygen poisoning on the succinoxidase system of insects, using *Drosophila melanogaster* as the experimental organism.

For more challenging study, a curious student should investigate other designs for respirometers and, employing a titration-type device, repeat the basic project for a comparison of results obtained by different techniques. A more advanced student should pursue the design and execution of a project in which he determines, by bioassay, the effectiveness of a series of teacher-prepared "unknown" chemicals as they act to inhibit or enhance the rate of animal or microbial metabolism.

Project—HABITAT AND NITROGEN METABOLISM

MOTIVATION

Some metabolic processes are amazingly geared to factors of the external environment. When converting amino acids to glucose by deamination, aquatic animals produce nitrogenous waste in the form of ammonia, whereas terrestrial forms tend toward the production of urea. Although ammonia is biochemically the more efficient form for the removal of nitrogen waste, it cannot be tolerated by delicate animal tissues and its distinctive odor guides scent-oriented predators unerringly to the excreters. It appears that only in animals living in water where the dilution factor is sufficient to negate the deleterious effects of the alkaline ammonia or in animals that are highly prolific can ammonia serve an organism's best interest. In all other animals, urea compounds, less efficient because they contain some carbon from food stores but without threat to the organism's safety and well-being, are the form taken by nitrogenous wastes resulting from the process of deamination.

MATERIALS REQUIRED

A project which investigates the form taken by nitrogenous wastes produced by animals in aquatic and terrestrial habitats will require:
    three dozen young tadpoles;
    a large aquarium;
    test tubes;
    2 ml. and 5 ml. pipettes;
    Nessler's reagent;
    1% gum ghatti;
    glacial acetic acid; and
    a methyl alcohol solution of xanthydrol.
For demonstration of colorimetry techniques and application, a colorimeter and/or photometer will also be needed.

DEVELOPMENT

Regular water-change and feeding routines should be established for 3-4 dozen young tadpoles being maintained in a laboratory aquarium, with one day per week designated for securing a water sample for testing. On the day preceding the sampling procedure, no food should be supplied after the daily water change. Once the scheduled 50 ml. sample has been siphoned from the tank, all daily routines should be resumed until 24 hours prior to the next scheduled sampling. This established routine should be repeated for the duration of the project and weekly water samples should be chemically tested for the presence of nitrogenous wastes from the organisms.

An ammonia test can be conducted simply by combining 6 drops of 1% gum ghatti with a 5 ml. water sample in a test tube, swirling gently, adding 3 ml. of Nessler's reagent, and continuing to swirl. The appearance of a yellow-brown color will indicate the presence of ammonia in the water sample.

A urea test can be easily performed by combining a 5 ml. water sample, 5 ml. of glacial acetic acid, and 0.5 ml. of a methyl alcohol solution of xanthydrol in a clean test tube. Settling out of large loose clumps after the contents of the tube have been vigorously shaken and allowed to settle for one hour indicates the presence of urea.

Results of the chemical tests should be recorded on a suitably prepared chart which also provides for notations concerning the corresponding developmental stage of the tadpole as evidenced by visual determination at the time of the water sampling. All routines established for feeding, water-change, fasting in clear water for 24 hours prior to sampling, and testing of aquarium water samples should be continued for a period of 3-4 months, or until young frogs have developed.

DISCUSSION AND INTERPRETATION

The data chart should be analyzed and the form of nitrogenous waste equated with the metamorphic stage of the organism and its natural habitat. The suitability of ammonia as a nitrogenous waste form for aquatic animals should be discussed in relation to the factors of safety and efficiency, and the production of urea compounds by terrestrial forms from the standpoint of the protection offered to the organism and as an adaptation to life on land. The enzyme patterns of nitrogen metabolism should be viewed as adaptive modifications for survival and indicators of evolutionary trends.

Consideration should be given to the biochemistry of protein digestion and deamination. The chemical testing procedures employed, the reasons for the animals' fast for a 24 hour period prior to sampling their nitrogenous wastes, and the limitations of a visual determination of ammonia concentrations should be reviewed and assessed. If available, a photometer or a colorimeter should be used to demonstrate the advantages of instrumentation in chemical analyses. (The color intensity of a Nesslerized sample is widely used as a reliable quantitative determination of ammonia production by organisms being studied.)

FURTHER STUDY

The basic project lends itself to an investigation of the form of nitrogenous waste produced by a variety of animals. Some students may wish to use an alternate experimental organism, such as the salamander, for the investigation. Others may be interested in a comparative study of unrelated animals representing both aquatic and terrestrial habitats. They should be encouraged to conduct a variation of the project, using goldfish, turtles, white mice, or other suitably maintained laboratory animals as the experimental organisms.

Some students should explore alternate techniques for the determination of the form of nitrogenous waste products that are produced by the experimental animals. They should investigate chemical procedures for the release of ammonia through decomposition of urea compounds produced and, employing these methods, design and conduct a project to be run in parallel with the basic study. Results thus obtained should be used for a comparative study and confirmation of the basic project findings.

Advanced students should design and conduct a project which explores the effect of ammonium sulfate solutions (in various small concentrations) on animal cells, using established cell cultures of mouse, monkey or chicken embryo cells grown *in vitro*.

Project—EFFECT OF TESTOSTERONE ON CO_2 PRODUCTION IN YOUNG CHICKS

MOTIVATION

The rate at which an animal engages in metabolism is influenced by its age, sex, size, body build, rate of growth, and muscular activity. In addition, certain metabolic-directed hormones which are specific for the control of biochemical reactions involving energy transformations serve to regulate the rate at which the vital body processes occur. It is through chemical reactions involving the oxidation of food to produce heat and energy that the rate of metabolism—the speed at which life activities proceed—can be measured.

Energy requirements for minimal functional activity can be determined by a direct measure of the heat produced by an animal engaging in no activity other than that of maintaining its high degree of organization; oxygen consumption and/or carbon dioxide production may be taken as an indication of the rate at which an organism's metabolic processes occur under varied conditions; and still another determination of the efficiency of the the body's chemical reactions, the amount of thyroid-produced protein-bound iodine (an index of the thyroid activity) may be separated from the inorganic iodine found in the blood stream to serve as an indicator of the rate at which metabolic processes are proceeding within the body.

MATERIALS REQUIRED

In a project which explores the effect of varying amounts of testosterone on the rate of metabolism in young chicks, the rate of carbon dioxide produced can be used as an index of the rate of metabolism. The following materials will be needed:

five groups of three young male chicks;
alcoholic solutions of aniline dyes:
methylene blue, malachite green, gentian violet,
fuchsin, eosin;
tuberculin syringes, 1 ml.;
hypodermic needles, #22 gauge;
bell jars;
250 ml. Erlenmeyer flasks;
an air pump;
assorted rubber stoppers, bent glass tubing, and plastic or
rubber tubing;
1000 ml. of 0.02N solution of NaOH with phenolphthalein
indicator;
a small animal weighing scale;
pasteurized testosterone propionate injection solutions:
100 micrograms/ml. in sesame oil,
500 micrograms/ml. in sesame oil,
1000 micrograms/ml. in sesame oil;
5000 micrograms/ml. in sesame oil; and
a control injection solution of pasteurized sesame oil.

DEVELOPMENT

Before beginning the project, students should be instructed in the techniques for proper care and handling of the young chicks to be used as experimental subjects. Also they should prepare the injection solutions needed and practice the proper technique for making subcutaneous injections. Five students, or groups, should then be assigned to work together for the total project. Each should mark the backs of three newly-hatched male chicks with an aniline dye solution, the color of which has been previously keyed to the assigned concentration of testosterone to be used for treatment.

On the first day of the project, all chicks in the first set should be weighed and subcutaneously injected with 0.1 ml. of a pasteurized solution of testosterone propionate at a concentration of 100 micrograms/ml. in sesame oil. This procedure should be repeated daily while chicks in the remaining four sets are treated similarly but with the previously designated concentrations of 500, 1000, 5000 and 0 micrograms/ml. testosterone in sesame oil.

After ten injections, during which time each subject will have received a total of 1 ml. of injectable material, a period of 3-4 days should be allowed for all of the material to be absorbed. During this waiting period, students should assemble apparatus for measuring the CO_2 output: connect, in series, an aquarium air pump, a 250 ml. Erlenmeyer flask fitted with inlet and outlet air tubes and containing 100 ml. of NaOH solution, bell-jar with two-hold rubber stopper for inlet and outlet air tubing, and another 250 ml. Erlenmeyer fitted with inlet and outlet air tubing and containing 100 ml. of distilled water with phenolphthalein indicator and enough 0.02N NaOH to produce a faint pink color.

When the proper waiting time has elapsed, sets of similarly injected chicks, as identified by the color code, should be placed under the bell jar for determination of CO_2 production. The time required to neutralize the NaOH by each set, depending on the total amount of testosterone administered by injection, should be determined and recorded. Similarly, the average size and weight gain and the output of CO_2, calculated on a per minute per gram of chick basis, for each of the five different situations should be recorded on the data chart.

DISCUSSION AND INTERPRETATION

The use of rate of CO_2 production as a technique for the determination of rate of metabolism should be discussed critically and evaluated. Chemical reactions involved in the addition of CO_2 to a solution of NaOH and the detection of the alkaline substance by a phenolphthalein indicator should be considered from the standpoint of the importance of employing chemical reactions for the detection of biochemical reactions.

A discussion of the project findings should not neglect to consider what meaning can be attached to the different rates at which CO_2 is produced by chicks having received different amounts of the testosterone injections. Differences in comb growth, crowing, increase in size and weight, and activity of the differentially testosterone-injected specimens should be noted and the role played by testosterone in whole body metabolism thoroughly researched and discussed in relation to the project. The metabolism-oriented functions of other hormonal secretions—thyroid, somatotropin, gonadotropic secretions, etc.—should also be discussed in this relation and, with sufficient motivation, serve as a springboard for initiating an extensive study of hormones and their regulatory effects on body processes.

On the basis of the project findings alone, both extreme and intermediary injections can be evaluated and the need for tempered amounts of "regulators" of metabolism indicated.

FURTHER STUDY

The project can be enlarged to include other amounts of testosterone and, through the inclusion of other concentrations such as 250, 2500, and 7500 micrograms/ml., a wider spectrum of dosage levels.

Variations of the project should be designed for the determination of rates of CO_2 production by insects, tadpoles, germinating and dormant seeds, yeasts, bacteria, and other micro and macro organisms. Bromthymol blue and phenol red as alternate indicators and $Ca(OH)_2$ and KOH as alternate alkalies should also be employed.

To check the validity of the results of the project, one student group may be assigned to research other techniques for the determination of metabolic rates. Employing the method using oxygen consumption as an index of the rate of metabolism, they should repeat the basic project and compare the results obtained by the two methods.

One or more students may wish to explore the effects of various concentrations of cortisone or thyroxin on animal metabolism. They should be encouraged to design a project which investigates these effects, using chicks or white mice as the experimental organisms. Other students who have developed great skill in dissection work may be given some assistance in the performance of a thyroidectomy on an experimental chicken or mouse which will be studied, with a control organism, in a comparison of metabolic rates.

These and countless other projects can be conducted by students since the study of metabolism is open-ended and provides fertile ground for investigations of original design.

Project—ENHANCEMENT OF BACTERIAL LUMINESCENCE

MOTIVATION

The familiar glow of the glowworm and the flashing light of fireflies are but two examples of the natural phenomenon known as bioluminescence; light emission is characteristic of a wide variety of organisms, from certain bacteria to some marine fishes. This cold light, varying from red to greenish-blue, is the result of a chemical reaction involving a luciferase enzyme and a non-protein substance specific to the species, and is produced as a secondary product of the organism's breakdown of organic compounds to secure energy for life activities. It may be that this is a carry-over from an early anaerobic ancestor who found in this device an effective method for drawing off and disposing of the accumulating oxygen with which it was otherwise unable to cope.

Employed by many kinds of present day organisms, mainly for some sort of recognition, communication, or attraction, this phenomenon of lower forms has recently been under exploitation by man: moistened dried Cypridinians yielded sufficient light for Japanese soldiers to read their maps yet remain undetected while engaging in jungle warfare in the South Pacific during World War II; fireflies and dinoflagellates are used advantageously for bioassay of ATP-containing compounds; research studies have indicated a close relationship of "living light" to other reactions in living cells, with implications for a new approach to the investigation of the fundamental processes involved in animal and human metabolism.

MATERIALS REQUIRED

Each student, or group, participating in the project which

investigates the effects of various salt concentrations on the amount of light emitted by a genetically luminescent bacterial species will need:

a stock culture of *Photobacterium fischeri;*
ten sterile culture tubes;
media ingredients;
  agar,
  peptone,
  sodium chloride, and
  distilled water;
an inoculating loop; and
a Bunsen burner or alcohol lamp.

An autoclave or pressure cooker and access to a darkened, room-temperature incubation zone will also be required.

DEVELOPMENT

The project, basically a comparative study of light emission by the genetically bioluminescent *Photobacterium fischeri* subcultured on media containing varying concentrations of sodium chloride, will require differential media. Each of six students groups should be assigned a different preparation, to be made up in sufficient quantity to supply the needs of all groups engaging in the project.

Basic media to be used for the control:

Agar	2.0 gm.
Beef extract	0.3 gm.
Peptone	0.5 gm.
Sodium chloride	3.0 gm.
Distilled water	100.0 ml.

Experimental media should be made up according to the same formula but with varying amounts of sodium chloride; 4.5, 4.0, 3.5, 2.5, 2.0, 1.5, 1.0, 0.5, and 0 gm. will give a good range for a basic study.

Each group should prepare media as assigned, mixing ingredients and heating gently, while stirring constantly, to the point of boiling. After boiling for one minute, agar should be allowed to cool to 45°C and, after being dispensed into culture tubes in 9 ml. quantities, autoclaved at 121°C under 15 pounds pressure for 15 minutes. Upon removal of the tubes from the autoclave, agar tubes, properly labeled to indicate the salt concentration, should be allowed to solidify on a slant board and refrigerated until needed.

Each student group should then assemble the materials needed for conducting the project study: stock culture of *Photobacterium fischeri,* ten agar slant tubes (each with a different concentration of sodium chloride), an inoculating loop, and an alcohol lamp or Bunsen burner. Using sterile technique, they should apply a heavy inoculation of the organism from the stock culture to each agar tube in the series. Tubes, properly labeled, should be incubated in a dark place at room temperature for 2-3 days and examined, in a dark room, for a visual determination of the degree of luminescence at each salt concentration. Using value judgments, consecutive numbers should be assigned to each subculture in ascending order of the degree of brightness discerned. An ingenious student might devise a luminescence meter, using a dark-lined mailing tube, 6 volt dry cell, photocell and galvanometer, by which light intensity readings can be obtained. Determined by either method, the results should be graphed, plotting brightness of illumination against salt concentration.

DISCUSSION AND INTERPRETATION

All data should be analyzed, conditions for the enhancement or inhibition of light emission determined, and other factors which might be expected to produce similar effects considered. The unusual phenomenon of light production by living systems should be thoroughly researched and discussed; the chemical aspects of light production as a side branch or by-product of some animal metabolism should be established; and the role played by aldehyde and luciferase in bacterial forms compared with the luciferin-luciferase action in fireflies, where a similar effect is produced.

The use of luminescence to the organisms of which it is characteristic should also be considered, and special attention should be given to the close relationship of animal luminosity to other chemical reactions occurring within their cells.

An interested student can demonstrate the phenomenon of bioluminescence by moistening a pinch of dried Cypridinians, previously crushed in an evaporating dish or on the back of his hand. The value of the light production here, and in the many other instances in which it occurs, should be emphasized as a tool for researchers in the study of some of the basic chemistry involved in living systems. Additionally, chemical formulas for luminescence produced by non-living substances can be investigated, demonstrated to the class, and compared, biochemically, with true bioluminescence.

FURTHER STUDY

Students should investigate further the factors affecting the luminescence of the photobacterium. Projects can be designed to determine the effects of temperature, other food components, oxygen concentration, antibiotics, long-chain aldehyde additives such as Dodecanal, drugs, and other inhibitors of cell respiration, and can employ suspensions of the bacteria in broth media as well as agar subcultures. Those wishing to study bacterial species from natural sources should investigate methods for the isolation of marine luminescent bacteria from salt water fish and, having isolated these forms, run a comparative study of light production by marine and non-marine forms.

Firefly luminescence also should be investigated. Advanced students should research procedures for ATP assay using luminescence produced by firefly tails and design a project which employs this technique to bioassay a group of substances which may contain ATP. A challenging study, such a project should include both quantitative and qualitative determinations.

Mechanically inclined students should be encouraged to devise accurate means for measuring the luminescence produced by living systems. Photocells and other devices should be designed for precise measurement and, using these and instruments for the determination of rate of respiration, a project should be designed and conducted to investigate the correlation of these two metabolic processes.

RECOMMENDED READING

Bender, A. Douglas. "A Consideration of Energetics in Biological Systems," *The American Biology Teacher*, (1965), 27, 1.

Boyer, Don R. "Hypoxia: Effects on Heart Rate and Respiration in the Snapping Turtle," *Science*, (1963), 140, 3568.

Brill, Winston J., E.A. Wolin, and R.S. Wolfe. "Anaerobic Formate Oxidation: A Ferredoxin-Dependent Reaction," *Science,* (1964), 144, 3636.

Bronk, J. Ramsey. "Thyroid Hormones: Control of Terminal Oxidation," *Science,* (1963), 141, 3583.

Consolazio, C. Frank, Robert E. Johnson, and Louis J. Pecora. *Physiological Measurements of Metabolic Functions in Man.* McGraw-Hill Book Company, New York, 1963.

Dorough, H. Wyman, Norman C. Leeling, and John E. Casida. "Nonhydrolytic Pathway in Metabolism of N-Methylcarbamate Insecticides," *Science,* (1963), 140, 3563.

Frank, O., H. Baker, H. Ziffer, S. Aaronson, S.H. Hutner, and C.M. Leevy. "Metabolic Deficiencies in Protozoa Induced by Thalidomide," *Science,* (1963), 139, 3550.

Harvey, E.N. *Bioluminescence.* Academic Press Inc., New York, 1952.

Jacobsen, A. and D. Gillespie. "Metabolic Events Occurring During Recovery from Prolonged Glucose Starvation in Escherichia coli," *Journal of Bacteriology,* (1968), 95, 3.

Lehninger, Albert L. "Energy Transformation in the Cell," *Scientific American,* (1960), 201, 5.

_____ . "How Cells Transform Energy," *Scientific American,* (1961), 204, 3.

McElroy, William D. *Cellular Physiology and Biochemistry.* Prentice-Hall, Inc., Englewood Cliffs, N.J., 1961.

McElroy, W.D. and H.H. Seliger. "Biological Luminescence," *Scientific American,* (1962), 207, 6.

Rasmussen, Howard. "The Parathyroid Hormone," *Scientific American,* (1961), 204, 4.

Schmidt-Nielsen, Knut. "The Physiology of the Camel," *Scientific American,* (1959), 201, 6.

Segal, S., H. Roth, and D. Bertoli. "Galactose Metabolism by Rat Liver Tissue," *Science,* (1963), 142, 3597.

Seliger, H.H. and W.D. McElroy. "Chemiluminescence of Firefly Luciferin without Enzyme," *Science,* (1962), 138, 3541.

Stumpf, Paul K. "ATP," *Scientific American,* (1953), 189, 4.

Verdin, J. "Simple Measurement of Respiration and Photosynthesis," *Turtox News,* (1963), 41, 234.

Weiss, William P. and Louis Sokoloff. "Reversal of Thyroxine-Induced Hypermetabolism by Pirmycin," *Science,* (1963), 140, 3573.

Wohlschlag, Donald E. "Metabolic Requirements for the Swimming Activity of Three Antaractic Fishes," *Science,* (1962), 137, 3535.

7

ANIMAL BEHAVIOR

What causes lemmings to seemingly run berserk in their final death throes, laboratory rats to learn a maze habit more readily than earthworms, birds to engage in migratory flights, or a paramecium to back up when it bumps into a snail? Why are the behavorial patterns of one animal species different from those of another? Is all behavior effective, or if not, can its characteristic pattern be improved or modified? These are some of the problems that are studied by ethologists and research physiologists and psychiatrists. Their research findings indicate that the behaviors of animals, from simplest forms to man, are no more perfect than are their physical structures; their behavior patterns, as exhibited, are simply the best that have evolved for the environmental conditions. They are finding, also, that under certain conditions the behavior pattern of an animal can be modified to give some advantage to both the individual and the species. It is here that important implications for man are indicated.

Project—RESPONSE OF GREEN HYDRA TO WHITE AND COLORED LIGHT

MOTIVATION

Even the simplest of animals are capable of responding to

136

specific factors of the environment; they may turn their entire bodies toward or away from outside stimuli as they respond to factors of heat, light, chemicals, touch and gravity. Equipped with only rudimentary structures, they engage in fundamentally stereotyped behavior patterns called *taxes* and, through responses which are highly predictable, gain a unique advantage in survival. The satisfactory adjustment of a simple animal to its environment results, in large measure, from a series of such behavioral responses; it is through taxes that an animal chooses appropriate food, selects a proper habitat, and avoids some of the environment's more hazardous elements.

MATERIALS REQUIRED

A project which investigates the response of green hydra to white and colored light can be conducted with simple and readily obtained materials. Each student or group will need:

a culture of *Chlorohydra viridissima;*
a glass culture bowl;
a 100 watt electric light bulb or other bright light source;
black construction paper;
colored cellophane disks:
    red, yellow, blue, green; and
a dissecting microscope.

DEVELOPMENT

Students should prepare from black construction paper a complete but removable covering for a culture bowl, thus rendering it light-proof. Then, five equally spaced windows, ¾ inch in diameter, should be cut or punched out of the paper base and a different color cellophane disk taped over each of four of the openings. The fifth window should be left uncovered for the admission of white light.

Twenty-five hydra from a thriving culture of *Chlorohydra viridissima* should be transferred, plus media, to the uncovered culture bowl where they should be left undisturbed for at least one day at 68° F in reduced light. After 24 hours, the paper covering should be placed over all surfaces and top of the bowl. The bowl should then be carefully placed on a support stand so that the bright light source is ten inches below the base of the bowl and all windows are so situated as to receive light of equal intensity.

After 24 hours the culture bowl should be carefully transferred to the stage of a dissecting microscope and the top paper cover removed. Leaving the side covers and five-window base cover in place, the number of hydra in each light area should be counted. With care being exercised not to disturb the hydra, the top paper cover should be replaced and the bowl returned to its support above the light source for another 24 hours. This procedure should be repeated for 4-5 days, with the number of hydra in each light area recorded on a chart and graphed for the various time periods.

DISCUSSION AND INTERPRETATION

The results of the project should be discussed and interpreted in terms of why the green hydra responds as it does to colored and white light and what adaptive value is gained by the exhibited taxis. The results should be correlated with photosynthesis of the green algae in the hydra and speculations made as to whether other organisms such as non-green hydra and the protist, Euglena, would be expected to exhibit the same responses.

Phototaxis should be viewed as a behavior pattern. The simplicity of the pattern should be noted and an analogy drawn to the relative simplicity of the organism making the response, with due consideration being given to the influence of the symbiotic zoochlorellae.

Other tactic responses should be investigated and the phototaxis, here illustrated, related to phototropism and photoperiodism, exhibited by certain other organisms, in a manner that brings about a comparison of the three patterns of behavior.

FURTHER STUDY

The basic project lends itself to many extensions including investigations into tactic responses to electricity, gravity, pH, drugs, vitamins, hormones, light intensity, chemicals, and heat. Planarians, paramecia, and both green and non-green hydra can be used as the experimental organisms.

Some students should investigate the *reflex* as another simple behavior pattern. Using *Hydra oligactis* as the experimental organism, a feeding response, normally reserved for live food, can be initiated through the stimulus of any glutathione-treated particle. Initiated by a fresh blood clot, this reflex behavior pattern makes a colorful and impressive demonstration for a student to perform before a class group.

Project—EFFECTS OF DRUGS ON WEB-SPINNING BEHAVIOR OF THE COMMON GARDEN SPIDER

MOTIVATION

Most animals come into existence fully equipped with the behavioral patterns characteristic of their own species. From their ancestors they inherit genes for these behaviors as well as for the structural features which permit their expression. Although greater skill may be developed through repeated performance, the innate behavior patterns are products of natural selection and involve no true learning of the type observed only in animals possessing more highly evolved nervous systems and brains. However, instinctive or learned, all effective behaviors offer survival advantage to the animals in which they are exhibited.

Recently, attention has been focused on agents which may act to interfere with established behavior patterns in man and animals. This is an important area of research study; investigations into the effects of drugs on animal behavior may reveal important insights into the potential usage of these substances to treat man's emotional and behavioral problems and, where modifications detrimental to effective behavioral patterns are observed to result from their use, to indicate clearly the advisability of man's avoidance of their indiscriminate use.

MATERIALS REQUIRED

The effects of drugs on the web-spinning behavior of the common garden spider can be investigated by individual or small groups of students equipped with:

common garden spiders of the genus *Argiope;*
an escape-proof box with fine wire screen cover;
a #25 gauge hypodermic needle;

a 1 cc. tuberculin syringe; and
an experimental drug, such as phenobarbitol.

DEVELOPMENT

Spiders to be used in the project should be maintained in the laboratory for several days before the experimental work begins. A relatively cool temperature and subdued light, a few leaves and twigs, and an occasional insect will help the spiders to adjust to the box enclosure, as observations of their activities will reveal. During this time, the characteristic pattern of the webs produced and the time required for this activity should be noted. Drawings of the web pattern should be made or, if desired, natural webs can be sprayed with a very fine mist of laquer from a spray can held at a distance from the web and transferred, with great care, to a piece of construction paper where they may be studied, intact, at a later date. During this period, also, students should practice feeding thirsty spiders a 1% glucose solution from the tip of a hypodermic needle attached to a tuberculin syringe.

After a few days, when spiders are fully acclimated to the box environment and students have developed some skill in the feeding technique, a phenobarbitol solution should be prepared by dissolving a phenobarbitol tablet in 25 ml. of 1% glucose. Using the technique previously developed for glucose feeding, students should administer one or more drops of the drug solutions to each experimental spider. Spiders should then be left undisturbed until their next webs have been spun, at which time observations and records of the web-pattern and time required for its production should be made and another dose of the drug administered. This entire procedure should be repeated 4-5 times, with observations of the time required for production and pattern of the web recorded for each drug dosage.

Similarly, the effects of other concentrations of phenobarbitol and alternate drugs such as benzedrine and chlorpromazine can be studied.

DISCUSSION AND INTERPRETATION

The records collected should be analyzed and the effects of phenobarbitol on both the web pattern and the time required for its production determined. If other drugs have been used also, a comparison of their effect should be made and the drugs classified on the basis of their effects on this behavior pattern.

On the basis of the project findings, students should discuss the practical value of using spiders to bioassay a new drug to determine its potential effect on human behavior and should evaluate the practice of a beekeeper who feeds her bees tranquilizers in sugar water so they will not sting when handled.

Insecticides which, in their final action, block the effect of the enzyme cholinesterase—essential for all nerve activity—and drugs such as LSD-25, where 1/4000 of an ounce causes hallucinations in humans, should be discussed from the standpoint of their mode of action in bringing about altered behavior patterns in the users.

FURTHER STUDY

Variations of the basic project can be designed for investigating the effects of alcohol, caffeine, cola drinks, and weak solutions of some insecticides on the web-spinning behavior of the garden spider.

Students wishing to determine the effects of drugs on other animals should design and conduct a project in which sowbugs, goldfish, or white mice treated with tranquilizers are compared on the basis of their maze performance with untreated specimens. An extension of this project pattern to include stimulants can be investigated by the same group or other students.

In still another project, students can plan a project in which the effects of common substances such as tea, coffee, aspirin and cola drinks on the maze performance of small animals or the web-spinning behavior of spiders is investigated for the purpose of classifying the substances as stimulants or depressants of nervous system activity.

Project—PHYSICAL OBSTACLES AND STRENGTH OF FOLLOWING-RESPONSE IN NEWLY HATCHED CHICKS

MOTIVATION

A brood of ducklings following the mother is both a familiar sight and a study in animal socialization. Capable of moving around when only a few hours old, the precocial birds respond quickly to the mother and follow her about in what proves to be a long-lasting behavioral pattern. It appears that the nervous systems of the newly hatched birds are pre-programmed to make the following-response to moving objects and, if imprinting is accomplished during a critical period, a permanent behavior-controlling impression results. Since the mother is usually the first moving object to be observed, the hatchlings learn to follow her in what amounts to a curious cross between an instinctive and a learned behavior pattern.

Surprisingly, other objects can also elicit the same response and a "mother substitute" may serve as the imprint object. Thus, hatchlings will show a decided preference for the imprinted object, be it natural mother or inanimate substitute.

MATERIALS REQUIRED

For a project which investigates the effect of physical obstacles in the path of young chicks which are being imprinted, the following will be needed:
20 newly hatched chicks;
a 20-compartment brood chamber;
a circular training runway;
a smooth-surfaced floor for the runway;
a rough-terrain floor for the runway;
alcoholic solutions of contrasting color aniline dyes:
methylene blue and fuchsin;
a colored electric light bulb or other imprint object; and
a live mother hen or other test object.

DEVELOPMENT

On the first day of hatching, all chicks should be kept in the incubator, maintained at 103° F, until they are strong enough to stand and their feathers have dried. Students should devote this day to preparing the pieces of equipment that will be used:
1. A brood chamber should be constructed and compart-
    mented into at least 20 chambers with walls high enough
    to prohibit chicks, when placed in their individual cells,
    from viewing anything other than the floor and side walls.

2. A circular training runway, with a diameter of about 18 inches, outside wall about three inches high and a smooth surfaced floor should be fashioned from heavy cardboard, plywood, or other suitable material.

3. An alternate, removable runway floor, with plaster-of-paris constructed low hurdles and other surface irregularities should be designed and constructed.

Training should begin on the second day. Twenty healthy chicks should be selected and divided into two groups—one to be trained on the smooth, the other on the rough flooring of the runway. They should be transferred quickly from the incubator to their individually assigned cells in the brood chamber and marked with a number from one to ten in the aniline dye color previously coded to the differential training they are to receive. All chicks designated for training on the smooth floor should be placed, one at a time, in the training runway for a ten minute training period. Each should be led, with the slowly moving colored electric light bulb, around the circumference of the runway for as great a distance as can be accomplished in the allotted time. As the training for each is completed, chicks should be returned promptly to their individual compartments in the brood chamber. Similarly, and for the same ten minute training period, each chick in the second group should be trained on the rough surface flooring and returned to its assigned compartment until the next scheduled training session.

Individual, ten minute training sessions should be repeated daily, for ten days, using the same procedures and conditions established for each group during the first training period and taking care to return chicks promptly to their isolation cells after each training session has been completed.

On the eleventh day, the following-responses of the chicks should be tested. Trained chicks should be placed, one at a time, in a large container with a live hen and the colored electric light bulb within clear view and at equal distances from the chick. The responses made by the chicks should be noted and recorded; within each group, a tally should be made of the number of following-responses made to the imprint object and to the living chicken. The number of responses to the imprint object made by those trained in the obstacle chamber should then be compared with those trained in the runway without obstacles.

DISCUSSION AND INTERPRETATION

The results of the project should be discussed in view of the decided preference that some chicks show for the imprint object and the effect of their having had to withstand the interference of obstacles in developing this attachment. The project findings might well be evaluated in terms of the principle, "the strength of imprinting is equal to the log of the effort expended by the animals during the imprinting period." Effects of other factors, such as age at which the imprinting is accomplished and the addition of sound to the inanimate imprint object as a means of enhancing the strength of the imprinting, also should be considered.

The work of Konrad Lorenz, whose imprinted goslings preferred him to their natural mother, and other researchers, who found that ducklings could be imprinted with a ticking alarm clock, should be investigated and discussed as they relate to the project. Other modified programmed behaviors, such as Pavlov's conditioning with dogs, should also be researched and the main differences between conditioning and imprinting clearly defined.

In light of the project findings, the relationships between early experience and socialization should be established; the effects of early experiences on the schooling of fishes and on courtship rituals of some birds and mammals should receive attention. Most importantly, the interest of the psychiatrist in the phenomenon of imprinting should be considered. How research in this area may reveal important insights into the mechanism of early experience and how these experiences may exert influences on human adult behavior are topics which stimulate great interest among students; they should be researched in the available literature and discussed in depth.

FURTHER STUDY

Students wishing to investigate the effect of age on the strength of imprinting should be encouraged to design and conduct a project in which the initial imprinting experience is offered at ages of 5, 10, 15, 24, 36 and 72 hours. Using young chicks or ducklings as the experimental organisms, the project should seek to discover the critical period for maximum effectiveness of imprinting and the age at which imprintability begins to decline.

For a more challenging study, students should design a project which seeks to discover if imprinting is limited to visual contact. Others may investigate, via controlled experimentation, the reported irreversibility of imprinted behavior patterns in a project which employs precocial birds as the experimental subjects.

Organisms other than birds should also be considered for use in studies of the following-response and other imprinted behavior. After developing proper techniques for handling young insects, guppies, and other fish, students should design projects which involve these organisms in investigations of the effects of early experience on socialization and adult behavior, with applications to the higher animals, and man, as the ultimate goal.

Project—MOTIVATION AND LEARNING

MOTIVATION

Professional and self-made nutritionists have long extolled the value of fish and other sources of high quality protein as "brain food." Aldous Huxley's *Brave New World* projects man's employment of biological controls as a means of producing individuals of desired intelligence levels. Science fiction writer Daniel Keyes spins a fascinating and convincing tale of Charlie, the mentally retarded adult, who is constantly beaten at maze puzzle solutions by the mouse Algernon until he too becomes a mental genius by virtue of the same brain surgery to which the mouse had been subjected. Researchers at the University of Michigan are currently studying human experimental subjects in an investigation into the effectiveness of Cylert, a magnesium pomoline preparation reported to have improved the capacity for intelligent behavior and the speed and retention of learning in similarly treated rats. Some of the above are instances of pure fiction while others represent true scientific endeavor, but in each is revealed man's concern and preoccupation with the thought of improving his intelligent and effective behavior patterns. In all, he seems determined to come to a better understanding of the mechanisms involved in his superior intelligence and to protect his vested interest in it.

MATERIALS REQUIRED

To engage in a project which investigates the effectiveness of motivation on the rate and permanence of learning a maze pattern, students will need:

12 white rats which are littermates;
a small animal maze;
aluminum foil;
dry cells; and
a stop watch.

DEVELOPMENT

Twelve young white rats from the same litter should be secured and maintained under conditions of extreme consideration and good care in the laboratory. During their adjustment period, attention should be given to the construction of a small animal maze from wood, plexiglass, or other suitable material. While any conventional maze pattern may be followed, the maze constructed should have six inch high walls, four inch wide paths, and a window screen covering which fits securely over the entire frame. It should include an entry way and, at some point, a goal box with comfortable nesting materials.

On the first training day, each rat should be given an introduction to the maze. One by one, they should be placed before the entry and, in the uncovered maze, led by hand along the paths leading to the comfort goal box. This procedure should be repeated several times and then, with each rat placed in the maze, one at a time and with all openings closed and the screening placed over the top, the time for each to reach the goal box should be determined and recorded.

As the rats are returned to their cages, they should be marked for identification as members of the Control Group or the Experimental Group and, within each group, with a number, one to six, corresponding to that assigned when the initial maze performance was timed.

On the second day, each member of the Control Group should again be placed in the maze entry and, with all conditions identical to those of the first day's test, timed as it finds its way to the comfort box. The floor and inside walls of the maze, except for the comfort box, should then be lined with aluminum foil and dry cells, with wires attached, made ready for use. The Experimental Group should then be tested; while each member is in the maze box, electricity, not exceeding six volts, should be applied from the dry cells. The time required for each experimental rat to reach "comfort" should be determined and all data for both groups recorded on a suitably constructed chart.

The procedure established on the second day should be repeated daily for 2-3 weeks, with daily records made of the time required for maze performance by each member of both the Control and Experimental groups. At the completion of each performance, rats should be returned to their respective cages and given gentle care and handling. For a comparison, graphs for the average time required by the Experimental Group to reach the goal should be plotted for each day that the project is run and, on the same axis, the average time for the Control Group to accomplish the same task should be plotted in a contrasting color.

DISCUSSION AND INTERPRETATION

The project data should be analyzed and the effect of motivation on learning the maze, as determined by the decrease in time required for maze performance from one trial in its solution to the next, discussed. Here, pain should be identified as the motivating drive and, in other instances, hunger, thirst, and sex as basic physiological forces that also exert a profound influence on animal behavior. The impact of these primary drives should be stressed.

The findings of researchers who have fed minced, trained planarians to untrained worms of the same kind should be investigated and their conclusions concerning the possibility of a transfer of learning reported to the class for discussion. Conditioned responses, as demonstrated by these trained worms and by Pavlov's dogs, should be identified as the basis of habit patterns and, very possibly, the cornerstone of all levels of learning.

Various levels of learned behavior should receive consideration, with students volunteering examples from their observations and experiences of conditioning, trial and error, observational and insight forms of learning. In each, the level of the learned behavior pattern should be equated with the evolution of the animal's nervous system, and thought should be given to the idea that morphology and behavior may have evolved together as adaptations to the environment.

FURTHER STUDY

If desired, the project may be extended for another several days. After the last run on the tenth day, a three day rest period should be allowed before the testing procedure is repeated again, this time for a determination of the degree of retention of learning by both experimental and control group animals. Other time lapse periods, such as five, eight, and ten days, may also be used.

The project also lends itself to many variations. For a comparative study, some students should be assigned the testing of alternate physiological drives, such as thirst and hunger and, using all available data, the relative strength and effectiveness in learning the maze determined for all drives tested. The factor of age, too, can be investigated. Employing young, middle-aged, and old rats, the average time required for each age group to solve the maze can be graphed on the same axis and used for a determination of the effect of age of rats on the rate of learning.

Highly motivated students should design projects in which the effects of nutrients, vitamins, hormones, drugs, antibiotics, magnetic fields, temperature and hyperbaric oxygen on the rate of learning are investigated. Maze performance or other suitable tests of learning can be employed, using rats, chicks, hamsters, or guinea pigs as the experimental organisms. For a more challenging study, students should investigate substances reputed to have the potential for improving the retention of learning. In this regard, special extracts, such as yeast RNA, should be checked for their effectiveness in improving the memories of rats previously trained in the basic project.

The technique of conditioning should be employed for the training of a variety of animals, from paramecia to planarians, earthworms, snails, and goldfish. Projects should be designed for testing the effects of punishment and rewards on the rate of learning an appropriately constructed "T" maze for each. As with the higher organisms tested, a 2-3 day rest period should elapse before re-testing and a comparison of the "before" and "after" rates taken as an index of the memory powers of the animals trained under different conditions. In a comparison of all results, the capacity for and the permanence of conditioning can be related to the level of nervous system development of the various organisms tested.

RECOMMENDED READING

Agranoff, Bernard W. "Memory and Protein Synthesis," *Scientific American*, (1967), 216, 6.

Bandura, Albert. "Behavioral Psychotherapy," *Scientific American*, (1967), 216, 3.

Best, Jay Boyd. "Protopsychology," *Scientific American*, (1963), 208, 2.

Boycott, Brian B. "Learning in the Octopus," *Scientific American*, (1965), 212, 3.

Cofoid, Dianne A. and Werner K. Honig. "Stimulus Generalization of Imprinting," *Science*, (1961), 134, 3491.

Cook, L., A.B. Davidson, D.J. Davis, H. Green, and E. Fellows. "Ribonucleic Acid: Effect on Conditioned Behavior in Rats," *Science*, (July 19, 1963), 141.

Dethier, V.G. and E. Stellar. *Animal Behavior*. Prentice-Hall, Inc., Englewood Cliffs, N.J., 1964.

Dilger, William C. "The Behavior of Lovebirds," *Scientific American*, (1962), 206, 1.

Eibl-Elbesfeldt, Irenaus. "The Fighting Behavior of Animals," *Scientific American*, (1961), 205, 6.

Farnsworth, Norman R. "Hallucinogenic Plants," *Science*, (Dec. 6, 1968), 162.

Flexner, J.B., L.B. Flexner, and E. Stellar. "Memory in Mice as Affected by Intracerebral Puromycin," *Science*, (July 5, 1963), 141.

Franchina, J.J. and M.H. Moore. "Strychnine and the Inhibition of Previous Performance," *Science*, (May 24, 1968), 160.

Fuller, J.L. *Motivation: A Biological Perspective*. Random House, New York, 1962.

Galen, Donald F. "Culturing Methods for Hydra," *American Biology Teacher*, (1969), 31, 3.

Geller, Irving. "Conditioned Anxiety and Punishment Effects on Operant Behavior of Goldfish," *Science*, (July 26, 1963), 141.

Gibson, E.J., T.J. Tighe, and R.D. Walk. "Behavior of Light- and Dark-Reared Rats on a Visual Cliff," *Science*, (1957), 126, 3262.

Gilbert, Perry W. "The Behavior of Sharks," *Scientific American*, (1962), 207, 1.

Hess, Eckhard H. "Imprinting in Animals," *Scientific American*, (1958), 198, 3.

Hingtgen, J.N. and M.H. Aprison. "Behavioral Response Rates in Pigeons: Effect of Methyl-m-tyrosine," *Science*, (July 12, 1963), 141.

Huxley, Aldous. *Brave New World.* The Modern Library, Random House, New York, 1932.

Jennings, H.S. *Behavior of the Lower Organisms.* Indiana University Press, Bloomington, Indiana, 1962.

Karsh, Eileen B. "Changes in Intensity of Punishment: Effect on Running Behavior in Rats," *Science,* (June 7, 1963), 140.

Keyes, Daniel. *Flowers for Algernon.* Bantam Books, New York, 1966.

Klopfer, Peter H. and Jack P. Hailman. *An Introduction to Animal Behavior.* Prentice-Hall, Inc., Englewood Cliffs, N.J., 1967.

Lewis, D.J. and H.E. Adams. "Retrograde Amnesia from Conditioned Competing Responses," *Science,* (Aug. 9, 1963), 141.

Lieberman, Michael W. "Early Developmental Stress and Later Behavior," *Science,* (Aug. 30, 1963), 141.

Lorenz, Konrad Z. "The Evolution of Behavior," *Scientific American,* (1958), 199, 6.

———— *On Aggression.* Harcourt, Brace & World, Inc., New York, 1966.

Nichols, John R. "How Opiates Change Behavior," *Scientific American,* (1965), 212, 2.

Polt, J.M. and E.H. Hess. "Following and Imprinting: Effects of Light and Social Experience," *Science,* (1964), 143, 3611.

Pribram, Karl H. "The Neurophysiology of Remembering," *Scientific American,* (1969), 220, 1.

Rensch, Bernhard. "The Intelligence of Elephants," *Scientific American,* (1957), 196, 2.

Roe, Anne and George Gaylord Simpson. *Behavior and Evolution.* Yale University Press, New Haven, Conn., 1967.

Rushforth, N.B., A.L. Burnett, and R. Maynard. "Behavior in Hydra: Contraction Responses of *Hydra pirardi* to Mechanical and Light Stimuli," *Science,* (Feb 1963), 139.

Russell, Claire and W.M.S. Russell. *Human Behavior.* Little, Brown and Co., Boston, Mass., 1961.

Sackett, Gene P., Patricia Keith-Lee, and Robert Treat. "Food vs. Perceptual Complexity as Rewards for Rats Previously Subjected to Sensory Deprivation," *Science,* (Aug. 30, 1963), 141.

Savory, Theodore H. "Spider Webs," *Scientific American,* (1960), 202, 4.

Schmidt-Nielson, Knut. *Animal Physiology.* Prentice-Hall, Inc., Englewood Cliffs, N.J., 1964.

Scott, John Paul. *Animal Behavior.* Doubleday & Co., Inc., Garden City, New York, 1963.

Skinner, B.F. "How to Teach Animals," *Scientific American,* (1951), 185, 6.

Thompson, William R. and Richard A. Dubanoski. "Early Arousal and Imprinting in Chicks," *Science*, (1964), 143, 3611.

Tinbergen, Niko. *Animal Behavior*. Time-Life Nature Library, New York, 1966.

_____ . "The Curious Behavior of the Stickleback," *Scientific American*, (1952), 187, 6.

Warden, Carl J. "Animal Intelligence," *Scientific American*, (1951), 184, 6.

Wecker, Stanley C. "Habitat Selection," *Scientific American*, (1964), 211, 4.

Weiss, Bernard. "Drugs and Behavior," *Science*, (June 7, 1963), 140.

Wesley, Frank. "Demonstrations in Animal Learning," *The American Biology Teacher*, (1966), 28, 4.

Wilson, Edward O. "Pheromones," *Scientific American*, (1963), 208, 5.

8

PLANT PHYSIOLOGY

One of man's primary concerns is with the well-being of the earth's green plants; he is completely dependent upon the basic food which they alone can synthesize for his sustenance and, in view of an impending population explosion, there is a danger that his needs may some day soon exceed the plants' capacity to adequately provide. Additionally, man must assume the responsibility for the continued existence and success of plant varieties which he has developed, for they are devoid of the advantages enjoyed by those which are products of a natural selection.

Through expanded knowledge gained via studies involving plant physiology and biochemistry, plant researchers can gain important insights into the mechanisms by which our plants can be propagated and maintained for optimal growth and production and, through wise practices based on these findings, man can protect his vested interest in green plants and endeavor to insure the very existence of future generations of mankind. The words of the *Bible,* "All flesh is grass," have great meaning for all times.

Project—EFFECTS OF PHOTOPERIOD ON FLOWERING IN CHRYSANTHEMUMS

MOTIVATION

Knowledge of the mechanisms involved in photoperiodism is a

152

boon to both commercial and home plant growers. By regulating the length of alternating periods of light and darkness, the production of one or more extra crops per year can be brought about and some seasonal plants can be induced to flower whenever desired.

The plant flowering and reproductive responses are genetically based and are characterized by a differential time-measuring feature, with orchids, wheat and snapdragons requiring long light exposures, while violets, poinsettias and soybeans respond to light periods of short duration and coleus, buckwheat and dandelions appear to be relatively neutral. Although the intricate details of the phenomenon are not yet fully understood, it appears that *phytochrome,* a light-sensitive pigment, triggers a photoperiodic response, that alternating light and dark periods of proper length cause the production of a hypothetical flowering hormone called *florigen,* and that any photo-reaction which interupts a period of darkness exerts an influence not only on fruit and flower formation but also affects seed germination, the lengthening of stems, and the production of plant pigments.

MATERIALS REQUIRED

To conduct a project in which the effect of the photo-period on plant flowering is investigated, the following will be needed:
   six mature, healthy chrysanthemum plants nearing the stage
      commonly referred to as "ripeness to flower;"
   two automatically controlled Gro-Lux fluorescent lamps; and
   two controlled growth chambers.

DEVELOPMENT

Prior to the time of beginning the project, the chrysanthemum plants should be grown under a photoperiod that maintains a healthy vegetative condition and produces plants of equal size and stage of maturity.

Students engaging in the project should place three such plants under a controlled photoperiod of 8 hours of light and 16 hours of darkness, and an identical group of three under a 16 hour light—8 hour dark regime. All plants should be cared for daily while maintaining the established photoperiod, and observations should be made for evidence of the onset of flowering. The project should be allowed to run for a period of 3-4 weeks, during which time accurate records should be kept for each plant being grown under each photoperiod. From the records, the effects of photoperiod on flowering should be determined.

If several students or groups engage in the project, other plants such as asters, begonias and violets can also be used for a comparative study.

DISCUSSION AND INTERPRETATION

On the basis of the project findings, students should·classify the chrysanthemum and other plants studied as short-day or long-day plants. After researching the literature for information concerning other photoperiod-influenced plant characteristics, they should discuss the formation of bulbs and tubers, the coloration of stems and leaves, and the importance of the proper photoperiod for plants making these responses.

The practical applications of photoperiod control should also be explored; the florist's use of artificial light to inhibit flowering of poinsettias and certain other plants until close to the date desired when, through increasing periods of darkness, they are brought to full bloom for peak sales periods, and the production of some highly seasonal flowers throughout the year by proper control of periods of light and darkness should be evaluated. The role of phytochrome in triggering the photoperiodic response should be established and the characteristic photoperiod of a given strain of plants linked to its genetic make-up.

The photoperiodic response should be discussed as a manifestation of photochemical control observed to occur in most of the higher plants and should be distinguished from other light-effected plant phenomena. In this respect, student observations of window plants leaning toward the light as they seem always to be looking out of the window will serve to demonstrate phototropism, which should be identified as a phenomenon uniquely different from that of photoperiodism investigated in the project.

FURTHER STUDY

The basic project can be extended to include a wide range of plant species and/or additional photoperiods. Studies in which gardenia, nasturtium, geranium, xanthium, snapdragon, alfalfa and radish plants are exposed to 12:12, 6:18 and 18:6 hour light regimes, as well as to conditions of constant light and constant darkness, should be planned for a determination of the photoperiod which is most and least suitable for the flowering of each.

Students wishing to investigate the effects of photoperiod on other plant processes might well design a project in which the reproductive rates of *Lemna minor,* grown in a balanced salt solution under various photoperiods, are determined and compared. Others should explore the effects of different wavelengths of light on the action of phytochrome; they should plan a project in which the growth of stems of common garden pea plants, exposed alternatively to red and far-red light, is compared to that resulting from exposure to red light only. This project, when undertaken, should be presented to the entire class for discussion as it serves to demonstrate the inhibition of growth by red light and the reversal of this inhibition by the far-red.

Motivated students should be encouraged to plan projects of original design and, where possible, to include a consideration of the effects of photoperiods on animal as well as plant forms of life.

Project—THE EFFECTS OF INDOLEACETIC ACID ON PHOTOTROPISM

MOTIVATION

The practice of applying a spray of very dilute indoleacetic acid to the unfertilized blossoms of tomato plants induces the development of seedless fruit by parthenocarpic means not unlike that observed to occur as a natural phenomenon in the production of seedless grapes. This and other practical measures, such as the development and effective use of 2,4-D selective herbicides and commercial compounds for the enhancement of the rooting capacities of certain plant cuttings, have grown out of research investigations into the identification and effects of regulators of plant growth. Additionally, the effects produced by the natural substance have been duplicated successfully by some synthetically produced chemical compounds. From whatever source, these substances are known as *auxins* and act to modify a variety of plant growth processes; differential growth, apical dominance, leaf abscission, flowering, growth of fruits, and proper orientation of a plant so that it is most advantageously adapted to its environment are regulated and controlled by their effects.

MATERIALS REQUIRED

The effects of indoleacetic acid on phototropism in cucumber plants can be studied as a class project, with small student groups, each pursuing one segment of the investigation, pooling results for a composite study. The total project can be completed in a period of 2-3 weeks and will require:

30 cucumber seeds;
30 250 ml. Erlenmeyer flasks;
30 foam plugs to fit the Erlenmeyers;
a 2% solution of sodium hypochlorite;
indoleacetic acid;
lanolin paste;
hydroponics culture medium;
toothpicks;
Gro-Lux fluorescent lamps or other light source; and
six seed germination chambers.

DEVELOPMENT

While cucumber seeds, previously soaked for 10 minutes in a 2% solution of sodium hypochlorite, are germinating on moist cotton or in germinating chambers, students should make the following preparations:

1. Foam plugs should be slit vertically, with a sharp razor blade, half way across from top to base.
2. Indoleacetic acid should be mixed with lanolin paste to yield final concentrations of 0, 1, 5, 10, 25, 50, 75, and 100 ppm.
3. A mineral-sufficient nutrient solution for hydroponic growth of seedlings should be prepared according to the procedure outlined in Chapter 5 and dispensed in 250 ml. Erlenmeyer flasks, leaving about 1 inch clearance for the insertion of a plastic foam plug.

When germinated, seedlings should be selected and transferred carefully to individual foam plugs where they should be placed on the inner flat surface between the slit sections and properly oriented—root growth directed downward and shoot growth flush with the top of the plug. The plug should then be folded around the seedling to secure its position and the entire assembly placed in the neck of a prepared flask so that the lower edge of the plug makes contact with the liquid in it.

After all seedlings have been so transplanted, they should be divided into ten groups of three and identified by proper labels to indicate the designated treatment:

1 -   three seedlings placed in a three-sided box so that light is unilateral

2 -   three seedlings grown in an open area with illumination of equal quality and quantity from all sides

3   three seedlings with plant shoot treated at level of upper surface of plastic plug with indoleacetic acid at a concentration of 1 ppm. in lanolin paste, applied with a clean toothpick and the entire assembly placed in an open area with equally distributed illumination

4 -   same as #3, but with IAA treatment at 5 ppm.

5 -   same   "   "   "   10 ppm.

6 -   same   "   "   "   25 ppm.

7 -   same   "   "   "   50 ppm.

8 -   same   "   "   "   75 ppm.

9 -   same   "   "   "   100 ppm.

#10 -   same   "   "   "   0 ppm.

Students should make daily measurements of the seedlings being grown under each condition and record observations on a data chart which provides for reporting indications of bending of stems as well as average daily growth for each group of seedlings over a 2-3 week period.

DISCUSSION AND INTERPRETATION

Students should pool the results of their assigned segments of the total study and, on the basis of the composite findings, determine the effect of indoleacetic acid on phototropism in plants. After researching the literature for background information, they should also discuss the relation of this substance to abscission, parthenocarpy, and circumnutation, and its employment in the regulation of plant activities. Spraying of fruit trees to retard ripening of fruit until it has attained a suitable size and color and to delay its natural fall; spraying of unfertilized flowers to produce seedless fruit; treatment of cuttings to promote rooting; and the use of 2, 4-D to promote an overgrowth leading to the death of broad-leaf plants, are all important practical applications which have grown from basic research into the nature and identification of this plant growth regulating substance that should be recognized.

An analysis of the procedures used in the project should also be made. Students should identify the control groups of seedlings and determine the need for their inclusion. They should also consider the advantages of using more than one subject for each growth situation and ponder the dangers that might arise from drawing conclusions from insufficient data.

FURTHER STUDY

Students may repeat the basic project, narrowing or extending the range of Indoleacetic acid concentrations or employing other plant seedlings as the experimental subjects. The synthetically produced growth substance, indolebutyric acid and naphthaleneacetic acid should also be investigated. Students should design and conduct projects in which their effects on phototropism are determined and compared with the naturally produced plant growth regulating substances.

Effects of auxins on other plant processes can also be explored; students should be encouraged to plan projects which seek to determine their effects on the rate of transpiration in tomato seedlings and the effects on photosynthesis in Chlorella. The role of plant growth substances in the determination of apical dominance can be demonstrated by applications of lanolin paste containing as little as 0.1% indoleacetic acid to the end of a growing shoot from which the apical bud has been removed. A highly motivated student might be asked to plan such a demonstration for class discussion.

The effectiveness of commercially available root promoting compounds, such as Auxan and Rootone, can be investigated in a project study which compares the roots developing on treated and untreated cuttings from geranium or begonia plants, and, if mature plants are available, students should be encouraged to plan projects which investigate the effects of weak solutions of one or more of the growth regulating substances on their leaves, flowers, and fruits. For another example of a practical application of man's knowledge about auxins, students should design a project in which a 2, 4-D preparation is applied, in recommended concentration, to a broad-leaf weed-infested patch of grass to determine its effectiveness as a selective herbicide.

Project—FUNGICIDES VS. PLANT PATHOGENS

MOTIVATION

No living things are without natural enemies, and plants are no exception. The commonly observed conditions of necrosis, defoliation, and/or abnormal tissue growth of various wild and cultivated plants are but symptoms of a diseased state which may be due to invasion by some parasitic fungus. In the performance of its life functions, the parasite brings injury, and sometimes death, to its host. While most effect a general drain on the host's resources, others, such as the *Caccomyces hiemalis* (which hydrolyzes amygdalin and causes a premature abscission formation and subsequent loss of leaves of the cherry tree) are specific in their mode of action and responsible for additional injury.

In most undisturbed natural conditions, host and parasite seem to have struck a balance, with the host and pathogen having evolved together. Barring any extreme change in one or the other, they are, and probably will continue to be, capable of co-existence. But natural conditions have all but disappeared from the scene and man's cultivation of plants has noticeably interfered with the balance. To meet the new challenge, plant pathologists are seeking temporary measures for combating parasites through the use of fungicides and the development of resistant host strains. The fact of evolution, of course, precludes all thought of ever developing plants which are permanently resistant to all parasites—man can merely speed up the evolution of the organisms, both hosts and pathogens.

MATERIALS REQUIRED

For a project which investigates the control of plant pathogens via the use of chemical substances, the following will be required for each student or group to use:

a supply of cucumber seeds;

sterile, neutral potting soil;

a 2% solution of sodium hypochlorite;

two flower pots or other growth chambers set in a shallow tray;

an Orthocide fungicide solution, 1.2 gm./500 ml. distilled water;

a culture of the plant pathogen, *Erwinia tracheiphila;*

two spray guns or atomizers; and

a greenhouse or other area of light and temperature conditions which are conducive to plant growth.

DEVELOPMENT

Equal numbers of cucumber seeds, previously soaked for ten minutes in a 2% solution of sodium hypochlorite or Clorox, should be planted, in duplicate, in seed flats, flower pots, or other growth chambers resting in shallow trays of water. Covered with clear plastic domes, they should be placed for 24 hours in an area of reduced light at a temperature of 76° F. Then, the seed flats should be transferred to daylight conditions at a temperature of 76° - 78° F where seedlings with established foliage should be allowed to develop. The seedlings, for all practical intents and purposes identical, should be labeled to designate one group as the experimental group, the other as the control. The experimental group should be sprayed with Orthocide solution, with care being taken to expose all plants to the same degree. In a similar manner, all seedlings in the control group should be sprayed with water.

After all plants in both groups have been allowed to dry, they should be equally sprayed with a water suspension of the culture of *Erwinia tracheiphila,* with special care being taken in the handling of the plant pathogen. Both experimental and control plants should be covered once more with their respective plastic domes and returned to conditions favorable for their continued growth. As a final step, all surfaces, equipment, and, most importantly, hands should be thoroughly washed until scrupulously clean and free of contamination.

Daily examination of the plants should be made and observations recorded with respect to growth, general appearance and other pertinent characteristics of the infected plants with and without benefit of a chemical protectant.

DISCUSSION AND INTERPRETATION

A comparison of the differentially treated infected plants should be apparent and both general and specific features should be cited. The pathogens should be examined microscopically and their appearance, motility, and gram morphology demonstrated. After researching the literature, the mode of action of *Erwinia tracheiphila* in attacking the host plant tracheae should be determined and, from the literature and project results, the use of selected chemicals to provide a protective barrier against such an attack should be firmly established as an ultimate advantage to the host plant.

Other methods of combating plant disease, such as the treatment of seeds before planting, should be considered and the characteristics of an effective and desirable fungicide determined. Non-mercurial and mercurial organic forms should be researched and the relative advantages noted. Orthocide should be evaluated on the basis of the project results and distinguished from the general group of chemicals known to be "fungistatic" in their action.

Plants closely related to the cucumber—cantaloupe, pumpkin, muskmelon and squash—should also be considered as possible targets of *Erwinia tracheiphila* attack and projects in which they serve as the experimental organisms planned for a determination of the general effectiveness of Orthocide.

Other plant pathogens and hosts should also be researched in the literature and should be discussed along with firsthand student observations of house and garden plants and ornamental and fruit trees with diseased conditions. *Pseudomonas tabaci, Corynebacterium insidiosum* and *Xanthomonas stewartii* should be considered from the economic as well as the botanic viewpoint.

A survey of the commercially available fungicides can be made and their labels and descriptive literature perused for specific recommendations. The controversial topic of spraying crops should not be neglected; students should be encouraged to research the topic thoroughly and to plan a spirited class debate on the question, "To spray or not to spray?"

FURTHER STUDY

Other fungicides can be used alternatively with Orthocide. Some students should make an evaluation of the effectiveness of a wide range of commercially available preparations such as Sanocide, Anticari and No Bunt against the pathogen *Xanthomonas stewartii,* using treated and untreated infected seedlings of a susceptible variety of sweet corn. Resistant strains of plants should also be compared with susceptible forms in a project study which exposes both to a known pathogen.

The effectiveness of fungicides when used alone and in combination with an insecticide should also be explored, using captan-dieldrin and thriam-dieldrin mixtures as well as the fungicide alone as a treatment for alfalfa seedlings infected with *Corynebacterium insidiosum.*

Additional projects should be designed to study *Erwinia amylovora* as it acts upon members of the Rosaceae family and *Phyomonas stewartii* on several varieties of sweet corn. Relationships of temperature, soil pH, and light intensity to the severity of wilting, necrosis, or other injury to an infected plant can also be investigated. Students should be encouraged to design projects involving these factors, using root cabbage plants infected with the microorganism *Plasmodiophora brassicae* or other host-pathogen combination for study.

Students wishing to explore the use of other agents for treatment of plant disease should consider the possibility of an effective natural relationship of antibiosis. They should cultivate colonies of a known plant pathogen on nutrient agar and introduce *Penicillium chrysogenum* or other antimicrobial organism in an attempt to locate an effective inhibitor and to compare the mode of action of the antibiotic substance produced with that observed in the case of the chemical fungicides.

For a very challenging study and practical application of the project study, each student should be encouraged to locate a diseased plant or tree in the locality, to diagnose the disease, and to plan a course of action for its treatment. A long range plan, this activity will prove to be most gratifying and rewarding to those students who can complete it successfully—to effect a cure for the diseased condition and to restore the plant to a state of good health and vigor.

Project—THE EFFECT OF GIBBERELLIC ACID ON UPTAKE OF P-32 BY LEMNA MINOR

MOTIVATION

The response made by some plants to applications of certain selected chemical substances may be readily observed; pea, bean, salvia, and other garden annuals show a marked elongation between nodes when treated with solutions of gibberellic acid and a noticeably shortened internodal growth when acted upon by Tributyl-2, 4-dichlorobenzylphosphonium chloride. The result is a correspondingly tall or short plant, either of which might be deemed more desirable, in a particular instance, than one of the normal height attained by a similar but untreated specimen.

Plant research scientists are exhibiting a decided interest in these striking results, particularly as they affect the growth-related processes of photosynthesis, respiration, and protein-synthesis. Toward this end, they are finding in radioisotopes a convenient and effective tool for indicating significant differences in physiological processes occurring within the mechanisms of the plants which, having received differential treatment, respond with resultant differential growth rates and patterns.

MATERIALS REQUIRED

Students can investigate the effects of gibberellic acid on the uptake of a mineral substance by water plants by engaging in a project study in which radioactive phosphorus and the common duckweed, *Lemna minor,* are used. For the project the following will be needed:

gibberellic acid;
a thriving culture of *Lemna minor;*
six culture bowls;

a balanced salt nutrient solution;

six 5-micorcurie units of P-32;

a Geiger counter;

autoradiography equipment: "no screen" X-ray film and X-ray exposure holder, fixer and developer;

radioactive warning signs;

rubber gloves; and

a darkroom equipped with a darkroom lamp.

DEVELOPMENT

Each student group should prepare a balanced salt nutrient solution, such as that recommended by Clark:

$$\text{Mix:} \quad 63 \quad \text{gm.} \quad CaH_2(PO_4)_2$$
$$809 \quad \text{gm.} \quad KNO_3$$
$$246 \quad \text{gm.} \quad MgSO_4 \bullet 7H_2O$$
$$2.7 \quad \text{gm.} \quad FeCl_3 \bullet 6H_2O$$

in distilled water to make one liter and adjust with:

1 M HCl and/or 1M KOH to pH 4.8.

The salt solution should be dispensed in 100 ml. quantities into each of six culture bowls or jars, with the remaining volume being retained for future replenishment of liquid lost by evaporation during the course of the project. Gibberellic acid should be added, differentially, to each 100 ml. of salt solution to produce a series of liquid media containing, respectively, gibberellic acid in 0, .01, 0.1, 1.0, 10 and 100 ppm. All culture bowls should be properly labeled to correspond with the concentrations of the additive.

After reviewing the instructions to be followed rigorously in the use of radioactive materials (wear rubber gloves, label and identify work areas through proper display of tags bearing the approved radiation symbol, thoroughly wash hands, utensils and all work surfaces with strong soap before and after work periods), students should very cautiously add five microcuries of P-32 to each culture bowl in the series. From the stock culture, they should select colonies of *Lemna minor* and transfer ten healthy specimens to each bowl, all of which should then be placed in an area with controlled artificial light for 16 hours per day.

Using a Geiger counter placed near each culture bowl, daily readings should be taken and recorded on a data chart for a period of 10-12 days. All data should be plotted on the same axis, with a different color being employed for each gibberellic acid concentration represented.

If desired, and if facilities and equipment are available, autoradiograms may be prepared on the final day of the project run; 3-4 colonies from each culture medium should be selected, measured with a mm. ruler or Bogusch measuring slide, and the exposure time in minutes for their autoradiographs calculated according to the formula

$$e = \frac{5 \times 10^5}{A}$$

where A is the activity in counts per minute of 1 cm.^{2} of specimen.

Exposure to X-ray film, according to prescribed procedures, in a dark room, will produce autoradiograms of the selected specimens and evidence, where indicated, of the uptake of P-32. All observations should be recorded on the data chart with notations made of the specific portion of the plant in which the phosphorus concentration appears to have occurred.

DISCUSSION AND INTERPRETATION

The data should be analyzed and a determination made concerning the effects of different concentration levels of gibberellic acid as it affects the uptake of phosphorus by the experimental plants. Evidence of accumulations of phosphorus in specialized plant organs, where detected, should be related to the metabolism of phosphorus by the plant, with correlations indicated, when and where applicable. In a practical manner, an evaluation of the project results should point up the implications for general and carefully regulated employment of this common agent which enhances plant growth.

The merits of employing radioactive tracer elements in research studies should be explored and students should be encouraged to offer information concerning specific instances in which they have seen the technique demonstrated and the precautionary measures practiced during the procedure. Those interested should plan additional projects, using a variety of different experimental organisms, other gibberellins, and, where possible, other tracers.

FURTHER STUDY

On the basis of the project findings, the range between concentrations of gibberellic acid found to produce no harmful results and those which proved harmful to the plants should be narrowed and an extension of the project planned to determine the tolerance point at which the substance does not interfere with the *Lemna minor's* ability to metabolize phosphorus. Variations of the project, employing other isotopes such as Na^{22}, I^{131}, and S^{35}, and other plants such as tomato, bean, and corn should also be planned for a determination of the influence exerted by Gibberellic acid on various plants to metabolize these elements. Other organisms may also be studied experimentally; a project can be designed to determine if a goldfish absorbs and concentrates phosphorus, sodium and/or iodine, and projects involving bacteria and yeasts should also be considered.

Advanced students may engage in further extensions of the project. Employing standard procedures, some should prepare chromatograms of the pigments extracted from the differentially treated duckweed fronds for a determination of the effects of varying concentrations of gibberellic acid on the chlorophyll pigments. Others, using similar techniques, may investigate the effects of ATP.

Some highly motivated students should research methods for extracting gibberellins from seeds of cucumbers, cantaloupes, beans, and peas, and, having prepared the extractions, test their effects on the metabolism of phosphorus, sodium and/or iodine by tomato, corn or bean plants grown hydroponically. The effectiveness of the growth promoting substances should be determined for various concentrations and compared with the commercially available gibberellic acid. Others might explore the effects of various concentrations of a known growth promoting substance on a genetically dwarf species, and, to reverse the situation, still other students may investigate the effects of various concentrations of Phosphon or other known plant growth inhibitor, following the pattern of the basic project design.

The effects of plant growth substances and of growth inhibitors on plant respiration should also be considered; students should be encouraged to design projects in which the effects of various concentrations of these substances on oxygen consumption and/or carbon dioxide production are determined, using tomato plants as the experimental subjects. Additionally, the plant physiological process of photosynthesis should be included; students should research methods for the determination of the rate of photosynthesis and, using Chlorella as the experimental organism grown in slightly acid water with various concentrations of a growth promoting or a growth inhibiting substance, discover the effects of these additives on the photosynthetic rate.

As an outgrowth of the basic project, the effects of chemicals on the chromosomes of plant cells may be explored. Students may engage in a project which investigates the cell abnormalities produced by colchicine treatment and, if sufficiently motivated, embark on an in-depth study of the effects of chemicals on cells grown *in vitro*.

Project—THE EFFECTS OF SOIL PH ON THE GROWTH OF LIMA BEAN PLANTS

MOTIVATION

As with all living things, green plants are extremely sensitive to factors of the environment; they react to conditions of temperature, quantity and quality of light, mineral content and abundance of water, and availability of oxygen to support their life processes. Any abrupt change in one or more of these vital factors may be reflected in essential physiological processes and, if severe, may result in the death of the plant.

Of special interest to researchers is the growth process. Through careful observations and accurate measurements of its rate when accompanied by some variable factor in a situation of controlled experimentation, optimal as well as detrimental conditions for a particular individual, variety or species can be determined. This bears important significance for home and commercial plant growers and has long range implications for all.

MATERIALS REQUIRED

A project which investigates the effects of soil pH on the growth of lima bean plants can be accomplished in a period of 3-4 weeks and will require for each student group:

18 lima bean seeds;
three plant growth containers;
sterile plant potting soil at pH 7.0;
garden variety pulverized lime;
screened peat moss;
a ½ inch hardware cloth soil sieve;
deionized water;
a 2% solution of sodium hypochlorite; and
plant growth measuring devices or auxometers.

DEVELOPMENT

While the lima bean seeds are being soaked in the sodium hypochlorite solution, soils to be used in the project should be prepared: one portion adjusted, by the addition of peat moss, to pH 5.0 - 5.5; a second, with lime, to pH 8.5 - 8.7; and a third, to be used as a control, being maintained at pH 7.0. Each potting mixture should then be sieved through the soil screen and used as the planting material for six seeds in a plant growth chamber appropriately labeled to indicate the specified soil pH. All plantings should be maintained in conditions of favorable temperature, light, and air and watered daily with deionized water to prevent any change in the established soil pH.

After the seeds have germinated, daily measurements of the height of each plant should be made and the average daily height for each group graphed on the same axis, using a different color for each soil pH value. Periodic observations concerning differences in leaf size and color, etiolation, sturdiness of stems, changes in pigmentation, and general appearance of plants grown in each soil should be recorded on a data chart for a 3-4 week period of time.

Three leaves, one selected as representative of that produced by each condition, should be surface marked with waterproof ink into 4 mm. squares and scale drawings of each made on graph paper. A daily record of accurate measurements of the numbered squares should then be kept and the growth rates of different leaf areas as well as differences in leaf growth produced by different soil pH values determined.

DISCUSSION AND INTERPRETATION

Generalizations can be drawn from an analysis of the graphs, charts, and general observations. Effects of soil pH on both general and localized plant growth should be determined from the data and students should be encouraged to evaluate the project findings objectively; they should look for possible sources of error and suggest reasons why any conclusion might not properly be considered valid.

The effects of both broader and narrower ranges of pH values of soil substances should be discussed, and other environmental factors, as they affect growth and other plant physiological processes, considered. The effect of light intensity on photosynthesis can be easily demonstrated by comparing the number of gas bubbles emanating from the needle-punctured stems of a series of sprigs of Elodea placed in a series of four glass graduated cylinders containing a 0.5% solution of sodium bicarbonate and exposed to the light of a 100 watt electric light bulb placed, differentially, at 6, 12, 18 and 24 inches from each cylinder. An able student should be asked to perform the demonstration and discuss it with the class.

FURTHER STUDY

Students will find other environmental factors also well suited for project studies concerning plant physiological processes. The basic project might be repeated, employing a wider or narrower range of soil pH values or a differential mineral content in the soil potting mixture as the experimental factor.

The biological effects of sound can also be investigated; students should research the use of audio generator, amplifier and oscilloscope set-ups and design a project which seeks to discover the effects of ultrasound on seed germination and plant growth, using corn, bean or radish seeds as the experimental organisms. Other wavelengths might also be investigated, with results from exposures to popular, classical, and rock-and-roll music forms compared for their relative value in promoting seed germination, plant growth, and speed and effectiveness in completing plant grafts.

Some students may be assigned to conduct an investigation of the relationship between electricity and plant growth. Using "B" radio batteries, electrical wires and battery clips, they can connect a grapefruit seedling into an electrical circuit and, using several plants for a comparative study, determine growth rates at different voltages. Other students may wish to investigate the effects of radiation on plant growth. They should obtain seeds which have received different amounts of gamma radiation and compare their germination and growth with that resulting from similar, but untreated seeds.

Advanced students should be encouraged to expand on the basic study and to conduct projects of original design in related areas. Projects which investigate the effects of various wave lengths of artificial light on the production of chlorophyll pigments, the effects of temperature and/or air currents on the rate of transpiration, and the growth patterns of plants maintained in a hyperbaric atmosphere present challenges in design and interpretation for the advanced and highly motivated students.

RECOMMENDED READING

Bhargave, S.C. "Inhibition of Flowering by Light in Short Day Plant *Salvia occipetals*," *Science,* (Jan. 1, 1965), 147.

Colby, S.R. and G.F. Warren. "Herbicides: Combination Enhances Selectivity," *Science,* (May 1, 1963), 141.

Fischer, R.A. "Stomatal Opening: Role of Potassium Uptake by Guard Cells," *Science,* (March 4, 1968), 160.

Fogg, G.E. *The Growth of Plants.* Penguin Books, Baltimore, 1963.

Friend, D.C. and V.A. Helson. *"Brassica campestris*: Floral Induction by One Long Day," *Science,* (Sept. 2, 1966), 153.

Galston, Arthur W. *The Green Plant.* Prentice-Hall, Inc., Englewood Cliffs, N.J., 1968.

Hendricks, Sterling B. "How Light Interacts with Living Matter," *Scientific American,* (September, 1968), 219, 3.

————— . "Metabolic Control of Timing," *Science,* (July 5, 1963), 141.

Hill, R. and C.P. Whittingham. *Photosynthesis.* Methuen and Co., Ltd., London, 1957.

Iwahori, S., S. Ben-Yehoshua, and J.M. Lyons. "Effect of 2-chloroethanephosphonic Acid on Tomato Fruit Development and Mutation," *BioScience,* (Jan. 1969), 19,1.

Krizek, D.T., W.J. McIlrath, and B.S. Vergara. "Photoperiodic Induction of Senescence in Xanthium Plants," *Science,* (Jan. 7, 1966), 151.

Lange, Clarence. (ed.) "Radioisotopes in Biological Research and Teaching," *The American Biology Teacher,* Special Issue, (1965), 27, 6.

Letham, David S. "Cytokinins and Their Relation to Other Phytohormones," *BioScience*, (1969), 19, 4.

Muir, Robert M. and Robert E. Yager. "Abscission," *Natural History*, LXVII (1958), 9.

Porter, R.D. and S.N. Wiemeyer. "Dieldrin and DDT: Effects on Sparrowhawk Eggshells and Reproduction," *Science*, (1969), 165, 3889.

Rabinowitch, E.I. "Photosynthesis," *Scientific American*, (1953), 189, 2.

――――. "Progress in Photosynthesis," *Scientific American*, (1953) 189, 5.

Salisbury, Frank B. "Plant Growth Substances," *Scientific American*, (1957), 196, 4.

Sax, K. *Standing Room Only*. Beacon Press, Inc., Boston, 1955.

Scott, T.K. and W.P. Jacobs. "Auxin in Coleus Stems: Limitation of Transport at Higher Concentrations," *Science*, (Feb. 15, 1963), 139.

Thomasson, W.N., W.E. Bolch, and J.F. Gamble. "Uptake of ^{134}Cs, ^{59}Fe, ^{85}Sr, and ^{185}W by Banana Plants and a Coconut Plant Following Foliar Application," *BioScience*, (1969), 19, 7.

Van Overbeek, J. "Plant Hormones and Regulators," *Science*, (1966), 152, 3723.

――――. "The Control of Plant Growth," *Scientific American*, (1968), 219, 1.

Voeller, Bruce R. "Gibberellins: Their Effect on Antheridium Formation in Fern Gametophytes," *Science*, (1964), 143, 3604.

Westing, Arthur H. "Geotropism: Its Orienting Force," *Science*, (1964), 144, 3624.

Winthrow, Robert B. (ed.) *Photoperiodism*. American Association for the Advancement of Science, Washington, D.C., 1959.

Yarwood, C.E. "Tillage and Plant Disease," *BioScience*, (1969), 18, 1.

9

TISSUE CULTURE

Although the life span of a chicken is normally about eight years, chick heart cells, cultivated *in vitro,* were maintained by Dr. Alexis Carrel for over 35 years and some human cell strains, obtained from individuals whose lives have since been ended, are still being maintained today. It appears that cells, when removed from a complex cellular organization within a living body and provided with proper nutritional and environmental conditions, are capable of experiencing a longevity not to be enjoyed if they are allowed to remain a part of the intact organism. Of even greater significance to the cytologist, the cells cultured in isolation can be observed directly as they engage in their life activities.

The methods of cell and tissue culturing are varied and play a major role in the advancement of many areas of biological and medical research: they may involve cell aggregates or tissue explants of plant, animal or human origin; they may be employed in studies of embryological development, cytology or metabolism; and they have important practical use in the routine screening of potential anti-cancer drugs, growth of viruses from which polio-myelitis and measles vaccines are made, and diagnosis of some malignant neoplasms. Widespread in use, cell and tissue culturing techniques constitute one of the researcher's most valuable tools.

172

Project—INDUCING SYNCHRONY IN CELL DIVISION

MOTIVATION

Researchers find in the embryonated hen's egg an excellent source of tissue for studies of animal cells grown *in vitro;* the embryonic tissues have a tremendous growth potential and, protected by the shell and shell membrane, are free from microbial contamination. When grown under favorable conditions and in suitable media, the cells can be observed, unencumbered, as they engage in growth and metabolic processes, as they react to changes in their physical and/or chemical environment, and as they respond to their neighboring cells. Exposed to an experimental condition, they may be compared with those in a control situation, and the experimental factor may be evaluated with a gain of some insight into its effect on a physiological process.

MATERIALS REQUIRED

A project which investigates the effects of a temperature change on the mitotic rate of chicken embryo heart cells grown *in vitro* can be accomplished in a period of one full week. In addition to a dozen embryonated hens' eggs for the preparation of a chicken embryo extract, the following materials will be needed for each student engaging in the project:

three 10-11 day old chick embryos;
five sterile 30 ml. tissue culture flasks;
sterile forceps and dissecting scissors;
two sterile sharp pointed scalpels;
sterile pipettes, pasteur capillary and 1, 5 and 10 ml.;
a sterile 50 ml. Erlenmeyer flask;
two sterile centrifuge tubes;
two sterile petri dishes;
20 ml. sterile commercially prepared Hanks balanced salt solution;
20 ml. sterile commercially prepared Growth Medium 199;
10 ml. sterile chicken embryo extract;
10 ml. sterile 0.25% trypsin solution;
10 ml. sterile 1.4% sodium bicarbonate solution; and
a microscope.

Extra sterile materials should be available to take care of emergency situations and an incubator, water bath, centrifuge, autoclave, and refrigerator with freezing unit will also be required.

DEVELOPMENT

Much advance planning and preparation will be required for the project. Embryonated eggs to be used for the embryo extract should be incubated several days before those to be ready for initiating the cell cultures on the first day of the project and all equipment to be used must be sterilized and ready. The techniques to be employed in the project should be clearly demonstrated and practice sessions for students should be included if at all possible.

Students may be assigned to individual and group tasks of assembling and sterilizing the needed equipment, with one adept group charged with the responsibility of preparing chicken embryo extract in sufficient quantity to accommodate the entire class. When the designated eggs have been incubated for 10-11 days, the embryos should be transferred to sterile petri dishes for the removal of heads and limbs. The bodies should be rapidly transferred to a sterile flask, frozen in a freezer, and then thawed and homogenized by passing through a syringe fitted with a wire mesh screen. An equal volume of Hanks solution should be added to the mince and the mixture frozen, thawed, and certrifuged for ten minutes. Shattered cells should be removed by a capillary pipette and the extract frozen and stored until needed in the project. Students should understand thoroughly the entire procedure and should practice sterile technique throughout the preparation of the embryo extract.

When the embryonated eggs to be used for establishing the cell cultures have reached an incubation age of 10-11 days, the project may be started. All needed materials should be assembled and, practicing sterile technique throughout the project, students should work quickly and carefully. With sterile forceps they should remove and transfer the embryos from three eggs to a sterile petri dish containing Hanks balanced salt solution and, with sterile instruments, excise the hearts and transfer them to fresh Hanks solution in a fresh petri dish. Here, using sharp pointed scalpels, the hearts should be cut into pieces about 1 mm. square and bathed in the balanced salt solution. After the excess salt solution has been removed by pipette, the tissue fragments should be rinsed in 10 ml. of embryo extract-saline solution and the whole mixture transferred, by pipette, from the petri dish to a sterile flask in which it should be allowed to settle. Then the supernate should be poured off and discarded, after which 5 ml. of trypsin solution should be added, by pipette, and the contents of the flask incubated for 30 minutes at 37° C. The mixture should then be homogenized by drawing up and expelling the mince several times through a sterile pipette and, after the addition of another 3 ml. of the trypsin solution, the refined homogenized mixture should be adjusted, with sodium bicarbonate solution, to pH 7.8 and incubated for 10 minutes at 37° C. The mixture should then be divided into two equal parts in sterile centrifuge tubes and, after centrifugation for ten minutes, the accumulations of red blood cells, as well as the supernate, removed. Cells should be re-suspended in 10 cc. of Growth Medium 199 per centrifuge tube and the material pooled before being dispensed, by pipette, in 4 ml. quantities in each of five 30 ml. tissue culture flasks.

The tissue culture flasks should be oriented, flat surface down, in the incubator and allowed to remain for 24 hours at 37° C. After the incubation period, they should be examined with a microscope for evidence of cell growth adhering to the flat surface and the location of a representative area of growth in each flask should be marked, with a wax pencil, on the outside surface. The flasks should then be numbered and incubated differentially:

\# 1 - at 45°C in a water bath for five minutes
\# 2 - at 45°C in a water bath for ten minutes
\# 3 - at 12°C in an ice bath for ten minutes
\# 4 - at 12°C in an ice bath for 20 minutes
\# 5 - at 37°C in an incubator.

After the designated treatments (#5 will serve as a control), flasks should be returned to the 37° C incubator for five days, being removed once every 24 hours for examination and cell counts within the uniformly marked areas. All data should be recorded on a suitably prepared chart and, using a different color for each culture, the daily cell counts graphed on the same axis.

DISCUSSION AND INTERPRETATION

At the end of the project run, the data chart and graph should be analyzed for indications of temperature-induced synchrony in cell division. Using the control culture as a point of reference, the effects of a temperature change on the rate of mitosis should be determined and any visual observations of cell abnormalities in morphology or growth pattern should be noted and discussed in relation to the early incubation time and temperature with which it seems to be associated.

Students should evaluate their skill in the employment of sterile technique and, in cases where contamination occurred, should trace, if possible, the point of its occurrence. They should reflect on the extreme precautionary measures to insure aseptic conditions (stressed throughout the project), on the value of the trysin treatments given to the tissue fragments, on the importance of using a chicken embryo extract, and on the nutritive and other values of the ingredients of Growth Medium 199.

The rate of avian cell division should be compared with the rate of bacterial cell division observed in previous microbiological studies and, after researching the relative mass of the two cell types, students should discuss the thought-provoking question of which type has the more efficient organization for cell reproduction.

The practical uses of tissue culturing methods should not be neglected. Students should research and report to the class for discussion their findings concerning current practices in both basic and applied medical research.

FURTHER STUDY

To broaden the scope of the project, some students might substitute kidney, liver, lung, or leg muscle tissue excised from the already sacrificed embryos used for a study of heart cells in the basic project. The economy feature here is further enhanced by the opportunity it affords for a comparative study to be made simultaneously with the basic study.

Students wishing to extend the project further should investigate methods for re-suspending and subculturing their cell cultures and then plan a study which explores the effects of various nutrients, drugs, chemicals or hormones which, when added, might be expected to accelerate or decelerate cell growth. For starters, they might use sterile colchicine in concentrations ranging from 0.01 - 1.0 mg./ml. of Hanks balanced salt solution as the chemical agent or investigate the effect of tri-iodothyronine on the rate of metabolism of cells grown *in vitro,* using visual observations, cell counts or rate of acidification of the media as an indication of the number of cells produced in the culture.

Alternate methods of measuring cell growth should also be investigated. Some students should repeat the basic project and, using a hemacytometer for determination of total cell count, compare the results obtained with those of the basic project study.

Other sources of cells and tissues should also be explored and students should be encouraged to design projects in which they investigate the nutritional requirements, effects of drugs, chemicals, ATP and radiation on cells derived from young fish, frog, mouse and insect larvae tissues. Regardless of the source of tissue or the experimental factor studied, students should be encouraged to plan and work thoughtfully and carefully and to engage in meaningful studies of original design.

Project—GROWTH OF SEEDLING ROOT FRAGMENTS CULTURED *IN VITRO*

MOTIVATION

The influence of chemical substances on the growth of specific plant structures is evidenced in many observations: gibberellic acid, applied to the stems of growing plants, produces plants which are abnormally tall; Amo-1618-treated pole bean plants become severely dwarfed; auxins at one concentration will stimulate root growth with no corresponding effect on stems, whereas at higher concentration levels it will promote the reverse effect; and phytokinins, used alone or in combination with auxins, act to stimulate the rate of cell division.

Researchers are finding their investigations into the reasons why chemicals produce these effects a challenging study and the technique of growing plant organs *in vitro* a major key to the solution of some basic problems in plant physiology.

MATERIALS REQUIRED

Aseptic conditions must be maintained when conducting a

project which investigates the effects of chemical substances on the elongation of plant roots isolated from the influence of other plant parts. Students engaging in such a project will need:

> radish seeds;
> two sterile petri dish germination chambers fitted with seed germination blotting paper disks;
> sterile distilled water;
> a 5% solution of sodium hypochlorite;
> coumarin;
> kinetin;
> six sterile petri dishes;
> sterile pipettes;
> sterile scalpels;
> sterile forceps; and
> chemicals for preparing a mineral solution growth medium:
>> $Ca(NO_3)_2 \cdot 4H_2O$
>> KNO_3
>> KCl
>> KH_2PO_4
>> $MgSO_4 \cdot 7H_2O$
>> ferric tartrate.

An autoclave and a dark incubation zone at 78° F will also be required.

DEVELOPMENT

Radish seeds to be used should be soaked in a 5% solution of sodium hypochlorite for ten minutes, then rinsed with sterile distilled water to remove all traces of the hypochlorite. The seeds should then be transferred (using alcohol-dipped, flamed forceps for the purpose) to sterile petri dish seed germination chambers containing sterile germination disks and 5 ml. of sterile distilled water. Each student or group should prepare two germination chambers, distributing 6-8 seeds evenly over the surface of each wetted blotter and placing the assembled chambers in a dark, undisturbed incubation zone at 78° F for the seeds to germinate.

The germination chambers should be checked daily and sterile distilled water should be replenished, as needed. During this time mineral growth solutions should be prepared, with student groups assigned to a specific segment of the preparation of a final solution to be used as either an experimental or control growth medium in the project.

> Stock solution #1 - 1.25 gm. $Ca(NO_3)_2$ • $4H_2O$
> 1.0 gm. KNO_3
> 0.875 gm. KCl
> 0.25 gm. KH_2PO_4
>
>> added, one at a time, to glass distilled water to make a total volume of 1 liter.
>
> Stock solution #2 - 0.2 gm. $MgSO_4$ • $7H_2O$ dissolved in 500 ml. of glass distilled water.
>
> Stock solution #3 - 0.1 gm. ferric tartrate dissolved in 500 ml. of glass distilled water.

The three stock solutions should then be combined: 80 ml. of solution #1, 100 ml. of solution #2, and 10 ml. of solution #3 with 810 ml. of glass distilled water to make 1 liter of mineral growth solution. This, dispensed in 100 ml. volumes in 250 ml. Erlenmeyer flasks, should be labeled and autoclaved at 121° C under 15 pounds pressure for 15 minutes.

Similarly, a second mineral solution, to which 2 mg. of coumarin has been added, should be similarly processed. And, in a like manner, a third preparation, the basic mineral solution plus 0.2 mg. of kinetin, should be made ready.

After autoclaving has been completed, all media should be cooled, poured in 25 ml. quantities into labeled sterile petri dishes, two per medium preparation, and left overnight as a sterility test. The next day, the young roots of the radish seedlings should be examined and six, selected on the basis of suitability of size and healthy appearance, should be chosen for experimentation. A 1 cm. length of each root tip should be cut with a sharp scalpel and transferred with forceps to individual prepared petri dishes that have passed the sterility test, two for each growth medium. Care should be taken to practice sterile technique throughout the procedure and to handle the petri dishes with care so that no splashing of media occurs on the sides and covers of the dishes.

By placing a plastic mm. ruler under the transparent petri dish, exact measurements of each root tip in each media should be taken and recorded on a suitable data chart. All labeled petri dishes should then be placed on a tray and returned to the incubation zone in a darkened area at 78° F.

The root tips should be examined and measured daily, with individual measurements recorded on the data chart and daily average lengths for roots grown in each medium plotted in a different color on the same axis. From the data chart and graph, maintained for a period of eight days, the rate of root growth in each growth medium should be determined.

DISCUSSION AND INTERPRETATION

The results obtained by all student groups engaging in the project should be pooled and the effects of the additives, coumarin and kinetin, on the growth of radish roots determined. The advantages of using the technique of tissue culturing for the study should be discussed; the benefits of an opportunity to observe the growth of roots isolated from the influences of other plant parts and the complexities of interactions of both inhibitors and enhancers of plant root growth which are known to be found normally in an intact plant should be considered.

The importance of practicing sterile technique throughout the project and the sodium hypochlorite treatment for seeds and subsequent washing with distilled water should be evaluated. In instances where contamination occurred, an attempt should be made to determine how it could have been prevented.

The data chart and graph should be analyzed and interpreted in terms of the rate of growth of the root segments cultured in each media and the effects of the additives in effecting an altered growth rate. The test substances used in the project should be classified on the basis of their influence on root growth and the chemical nature of coumarin and kinetin should be researched and compared with other chemical substances—indolebutyric acid, gibberellic acid and tributyl-2, 4-dichlorobenzylphosphonium chloride—known to exert some effects on plant growth.

FURTHER STUDY

Plans should be made to repeat the project using a wide range of concentrations of coumarin and kinetin and a variety of cells grown *in vitro* as the experimental specimens. Tomato, alfalfa, bean, pea and other seedlings should be used alternatively or in combination with radish seedlings for a comparative study of the effects of the chemical additives on root growth.

Brighter students should extend the project study by preparing squash slides of the root tips and, by microscopic examination, determining if the greater growth of a root segment, where observed, was due to an increase in cell size, cell number, or, possibly, a combination of the two. Others should plan a project to determine the effects of the same chemical substances on leaf tissue grown *in vitro*. Using the technique of measurement of leaf disks from primary pea plant leaves, grown for 24 hours in a sucrose solution containing the test agents, they should try to determine if the effects on root growth in the basic project are also observed in growth of leaves or if the action of coumarin and kinetin is specific for root tissue.

Students will find the tissue culture technique an interesting device for testing chemical substances. Inositol, salicylic aldehyde, sodium acetate, acetic anhydride, leucoanthocyanin and some teacher-prepared "unknowns" should be available for them to bioassay, using the techniques introduced in a project.

The discovery of natural substances having a kinin-like activity in plants offers a challenging exercise in research for advanced students. They should be encouraged to design projects which investigate the occurrence of kinin-like substances in coconut milk, corn endospern, fruit mashes, plant lectins and a wide array of other natural products, using a plant tissue culture technique for the identification of the kinins.

Project—ACTINOMYCIN D VS. REGENERATION OF HEAD TISSUE IN DUGESIA DOROTOCEPHALA

MOTIVATION

Observations of the *in vitro* development of a complete plant from a small tissue explant from the pith of the tobacco stem and of the rebuilding of a complete starfish, hydra or planarian body from but a small tissue fragment seem to bear evidence that the capacity for developing into an entire multicellular individual is inherent in every cell. However, problems involving cell specialization, the organization of a functional biological system, and proper nourishment for the developmental process frequently interfere with the realization of this potential.

It appears that many factors stimulate cell division and direct the development upon which regeneration of a lost part depends. Conversely, inhibitory factors operate to retard this process or to stop it completely once the desired stage has been reached. The researcher's concern with regeneration is to discover the factors which stimulate adult cells to develop into new structures and those which cause an inhibition of the rampant cell division that is so characteristic of cancer growth, and to apply these conditions, as needed, to man.

MATERIALS REQUIRED

A project which investigates the effects of Actinomycin D on the growth of a complete organism from a tissue fragment can be conducted in a period of 7-8 days and with the use of simple and readily obtained materials. Each student or group participating in the project will need:

12 *Dugesia dorotocephala,* black planarians;
five 8-inch culture bowls;
clear spring water;
sharp single-edge razor blades;
a 4-inch square glass plate;
Actinomycin D solution, 10 μg./ml. in spring water; and
a dissecting microscope.

DEVELOPMENT

Prior to beginning the project, students should maintain a culture of *Dugesia dorotocephala* in a culture bowl containing spring water at 70° F., with routine changes to fresh spring water made every 48 hours. Fresh beef liver should be provided during 30 minute feeding periods once a week, after which the organisms should be transferred to fresh spring water again.

To begin the project, 12 healthy specimens should be selected for study. Individually, they should be transferred to a glass plate and cut, horizontally, with a sharp razor blade, posterior to the pharyngeal region. The anterior regions should then be discarded or returned to the stock culture and the posterior fragments should be transferred, six to each of the previously prepared culture bowls: one containing Actinomycin D solutions (10 μg./ml. in spring water) and the other, serving as a control, containing fresh spring water only. Culture dishes, properly labeled, should be left in an undisturbed area at 70° F. for 48 hours.

After the specified time lapse, all specimens should be examined with a dissecting microscope for observation of any changes that have occurred. These should be recorded on a data chart and sketches should be drawn of the appearance of an individual specimen characteristic of the group in the experimental culture medium as well as one in the control group. Specimens should then be transferred to freshly prepared culture bowls containing spring water, properly labeled, and again set in an undisturbed area at 70° F. Examination of specimens should be made routinely over a period of 5-6 days, with records of observations and sketches made daily and transfers to fresh culture bowls of spring water made every 48 hours.

DISCUSSION AND INTERPRETATION

The growth of planaria fragments in the two cultures should be compared and, using the control group as a point of reference, the effect of Actinomycin D on the normal pattern of regeneration of new head tissue determined. Consideration should be given to the mode of action of the chemical substance as it inhibits the growth of tissue and the potential employment of this chemical to retard the growth of tumors.

The techniques used in the project also should be discussed; the suitability of spring water as a growth medium should be evaluated and the practice of suspending feeding periods for planarian fragments during the period of regeneration should be considered.

Students should research and report to the class on recent developments in the production of an entire aspen tree from a small plug of undifferentiated tissue grown *in vitro* and should discuss the significance of this technique in the study of nutritional needs of plants. The potential development of new formulations for fertilizers used in gardens, orchards and forests should be considered as a possible outgrowth of basic *in vitro* studies involving induced differentiation and growth enhancement of excised pith tissues from experimental plant stems.

FURTHER STUDY

Variations of the project, including other concentrations of Actinomycin D, other time exposures for the treatment of the experimental subjects, and time lapses of various duration between decapitation and Actinomycin treatment should be planned and run in parallel with the basic study for an in-depth coverage of the topic. Students wishing to determine the effects of Actinomycin D on other invertebrates should be encouraged to substitute hydra or fresh water sponge specimens for the planarians studied in the basic project.

The effects of other factors on regeneration of lost parts also should be explored. In addition to a wide array of antibiotics and other pharmaceuticals and chemicals, exposures to ultraviolet light and changes in temperature should be investigated, using arachnids, nematodes, fresh water oligochaetes and salamanders as the experimental organisms. And, after researching the methods reported to be successful in growing an entire carrot plant or aspen tree from a small plug of unspecialized tissue, students, using these *in vitro* techniques, should plan a project in which they bioassay a wide range of substances believed to exert an influence over the production of plant parts from unspecialized tissue.

RECOMMENDED READING

Collier, H.O.J. "Kinins," *Scientific American*, (1962), 207, 2.

Dipaolo, J.A. and C.E. Wenner. "Thalidomide: Effects on Ehrlich Ascites Tumor Cells *in Vitro*," *Science*. (Nov. 15, 1963), 142, 3594.

Farris, Vera King. "Molluscan Cells: Dissociation and Reaggregation," *Science*, (June 14, 1968), 160.

Flickinger, Reed A. "Cell Differentiation: Some Aspects of the Problem," *Science*, (August 16, 1963), 141.

Gibbs, J.L. and D.K. Dougall. "Growth of Single Plant Cells," *Science*, (Sept. 13, 1963), 141, 3585.

Harary, Isaac. "Heart Cells *in Vitro*," *Scientific American*, (1962), 206, 5.

Hirumi, H. and K. Maramorosch. "Insect Tissue Culture: Use of Blastokinetic Stage of Leafhopper Embryo," *Science*, (June 19, 1964), 144, 3625.

Knight, V.A., B.M. Jones, and P.C.T. Jones. "Inhibition of the Aggregation of

Dissociated Embryo-chick Fibroblast Cells by ATP," *Nature*, (June 1966), 210.

Kordan, Herbert A. "Starch Synthesis in Excised Lemon Fruit Tissue Growing *in Vitro,"Science*. (Nov. 15, 1963), 142, 3594.

Letham, David S. "Cytokinins and Their Relation to Other Phytohormones," *BioScience* (April 1969), 19, 4.

Levinson, C. and J.W. Green. "Response of Cultured Chick Heart Cells to Changes in Ionic Environment," *Journal of Cellular Biology*, (June 1966), 67.

Lewis, Hazel and Isaac Harary. "Calcium-Activated Adenosine Tri Phosphate Localization in Cultured Beating Heart Cells," *Science*, (July 5, 1963), 141.

Marks, Edwin P. and John P. Reinecke. "Regenerating Tissues from the Cockroach Leg: A System for Studying *in Vitro,"* *Science*, (Feb. 28, 1964), 143, 3609.

Mathes, Martin C. "Antimicrobial Substances from Aspen Tissue Grown *in Vitro,"Science*, (June 7, 1963), 140.

Merchant, D.J., R.H. Kahn, and W.H. Murphy Jr. *Handbook of Cell and Organ Culture*. Burgess Publishing Co., Minneapolis, 1960.

Paul, J.R. *Cell and Tissue Culture*. Williams and Wilkins, Baltimore, 1960.

Proffit, William R. and James L. Ackerman. "Fluoride: Its Effects on Two Parameters of Bone Growth in Organ Culture," *Science*, (1964), 145, 3635.

Puck, Theodore T. "Single Human Cells *in Vitro,"* *Scientific American*, (1957), 197, 2.

Rao, Potu N. "Mitotic Synchrony in Mammalian Cells Treated With Nitrous Oxide at High Pressure," *Science*, (1968), 160, 3829.

Ross, Russell. "Wound Healing," *Scientific American*, (1969), 220, 6.

Sasaki, M.S. and A. Norman. "Proliferation of Human Lymphocytes in Culture," *Nature*, (May, 1966), 210.

Silvan, James. "Life Under Glass," *Science World*, (1963), 14, 1.

———— . "Secrets of Plant Growth, "*Science World*, (1964), 16, 2.

Singer, Marcus. "The Regeneration of Body Parts," *Scientific American*, (1958), 199, 4.

Willmer, E.N. *Tissue Culture: The Growth and Differentiation of Normal Tissues in Artificial Media*. John Wiley and Sons, Inc., New York, 1958.

Winton, Lawson L. "Plantlets from Aspen Tissue Cultures," *Science*, (1968), 160, 3833.

Zeuthen, E. (ed.) *Synchrony in Cell Division and Growth*. John Wiley and Sons, Inc., New York, 1964.

10

CHALLENGES FOR TODAY
AND TOMORROW--
SOME UNSOLVED PROBLEMS

On the scientists' drawing boards for the immediate future and far beyond are ambitious plans for an improved technology that will make possible better, healthier, happier and longer lives for both present and future generations of mankind. Implied is a rate of technological growth far surpassing that which man has already experienced and found difficult to keep pace with in terms of understanding and coping with concomitant newly created problems and irreversible situations. In this respect, biologists have responsibilities that are without equal—they must seek opportunities for the improvement and control of life while directing technological advances away from practices that threaten to destroy its delicate balance and they must strive to ensure the continued existence of the natural world in a way that will enable man to live in harmony with it.

The biologists of tomorrow are among the high school students of today; it is they who will be faced with problems relating to survival. Hopefully they will be prepared to respond to opportunity as well as to react to need and will have the wisdom and foresight to evaluate the long range aspects of new innovations. To this end, they should be mindful of the time-honored counsel offered in Proverbs 4, 7: ". . .get wisdom: and with all thy getting, get understanding."

Project—THE SEARCH FOR NEW SUBSTANCES FOR THE IDENTIFICATION OF BLOOD ANTIGENS

MOTIVATION

There is real danger which accompanies the indiscriminate transfusion of blood from one person to another; the red blood cells of the donor may become agglutinated in the plasma of the recipient and the subsequent blockage of capillaries in the heart, brain, kidneys or other vital organs can cause death. Although members of the same species, humans do not share identical blood proteins and, if introduced, foreign substances often are countered by antibodies in a reaction not unlike that associated with a natural immunity or rejection of an organ transplant. It is imperative that only compatible bloods be used for transfusions and that the compatibility be determined by reliable methods.

Studies of major blood groupings—ABO, Rh, and MN—have provided deeper insights into the relationships between inherited factors which determine blood types and disease as well as having made possible the avoidance of unsuccessful transfusions. However, improved techniques are being sought for performing the mechanics involved in typing blood. Encouraged by the discovery of some bacterial and seed plant substances related to human blood antigens, researchers believe that these can be supplied in abundance and processed at low cost. What remains to be determined is the exact nature of substances which will be specific for each of the blood factors to be identified.

MATERIALS REQUIRED

To initiate the search for plant substances which can be employed for the identification of human blood types, the following will be needed:

    human blood samples: types A, B, AB and O;
    commercially prepared blood typing sera: anti-A and anti-B;
    sterile, disposable hemolets;
    blood typing or microscope depression slides;
    Alsever's solution;
    10X magnifying lenses;

an assortment of dried plant seeds;
a Waring blendor or mortar and pestle
normal saline solution; and
membrane sterile filter assemblies.

DEVELOPMENT

Several preliminary preparations should be made before beginning the project:

Alsever's solution: 46.6 gm. glucose,
                        9.6 gm. sodium citrate, and
                        5.04 gm. sodium chloride

should be dissolved in 1200 ml. of distilled water and adjusted to pH 6.1 with citric acid. The solution should then be dispensed in 2 ml. volumes in small, stoppered culture tubes, sterilized by filtration, and refrigerated until needed.

Normal saline solution: 58.45 gm. sodium chloride should be dissolved in distilled water to make 1 liter of solution.

Seed Extracts: seeds selected for testing should be sorted and processed; a small number of each type should be ground, separately, in a Waring blendor or with mortar and pestle, with the addition of distilled water for moistening, if necessary. The ground mash should then be combined with normal saline solution in a 1:10 ratio for the preparation of a 10% seed extract, sterilized by filtration, dispensed into small, sterile vials, labeled, and stored in a refrigerator until needed.

To insure a supply of blood representing each type within the ABO grouping, blood samples should be secured from several human donors and typed in accordance with standard procedures: use sterile disposable hemolet to secure blood sample from clean, alcohol-swabbed fingertip, discard first drop of blood and transfer the next two drops directly into a prepared tube of Alsever's solution, and, using standard anti-A and anti-B sera, determine the blood type by a simple agglutination test. The typing of the blood cells in solution should be conducted on clean slides, preferably mounted on a box with a built-in light source, and observed with the aid of a 10X magnifying lens for detection of clumping of cells when maintained at room temperature for a period of 2-5 minutes. Positive reactions in anti-sera will identify corresponding blood proteins which should be catalogued along with the number of the blood sample and the name of the donor. Blood samples should be refrigerated when not being used and fresh samples secured weekly.

Employing the same technique as described for the identification of blood antigens A and B with standard blood typing sera, each seed extract should be tested for corresponding agglutination of cells in the blood samples of known blood types in the stock collection. Reactions with each type blood should be recorded for each seed extract and the collection of extract samples widened at every opportunity to include all seed classifications.

DISCUSSION AND INTERPRETATION

An analysis of the data chart should be made to evaluate the degree of positive reactions and to identify plant seed extracts that produce an agglutination of cells in designated blood types as opposed to others which may be non-specific. Relationships between seeds giving positive reactions should be sought to help channel concerted efforts into investigations of specific plant groups, such as legumes, for more extensive research into classifications which may provide substances of value in this determination. Students should identify the "trial and error" method used in this project and evaluate its merits in the field of research.

Students should distinguish between agglutination and coagulation and assess the harm which might come to a recipient if incompatible bloods are mixed. Blood factors other than those of the ABO grouping should be discussed also and attention given to the probable existence of antigenically unique blood for each individual. The production of antibodies should be viewed as a body defense against the introduction of foreign antigens and the aspects of immune reactions which are posing some problems today should be researched and discussed. Rejection of skin grafts and organ transplants, autoimmunity contributing to some physical and mental disorders, and allergic reactions, methods of desensitization, and threat of death from anaphylaxis should receive attention in this regard. Also, blood protein analysis as a means for determining essential amino acid deficiencies should be considered as it contributes to greater efficiency in prescribing proper diet supplements, where indicated.

FURTHER STUDY

Students who experience some success in identifying seed extracts which are specific for blood factors tested might well extend the project to the point of practical application by producing these extracts in quantity. If extracts specific for A and B factors are located and identified, a project should be designed along similar lines for the investigation of seed extracts which might be used for the determination of Rh, M, and N blood factors. Testing of the plant substances should be performed in parallel with the standard testing sera to provide for a reliable basis for comparison.

To broaden the sampling of seeds tested, some students should enlist the aid of foreign exchange students, requesting them to send seeds from their homelands to be used in the on-going research project, while others investigate different sources of protein substances related to human blood antigens. They should be encouraged to design projects in which extracts obtained from bacteria, algae, lichens or other organic sources are tested for the presence of substances which can serve reliably and inexpensively in the typing of blood.

Some students may wish to pursue related research into immune reactions which often accompany skin and organ transplants. Employing careful techniques, they should plan and perform skin transplants on laboratory mice, having exposed the mice to varying doses of radiation, such as X rays, prior to performing the surgery. Only advanced students equipped with extensive background knowledge and perfected skills should attempt projects such as these in which the relation between blockage of the body's immune response and transplant surgery is investigated.

Project—THE SEARCH FOR NEW SOURCES OF FOOD

MOTIVATION

Those who have made population studies with yeasts, bacteria, and other organisms recognize the rapid growth rate currently being experienced by the human population as the exponential growth stage of a typical population growth curve. They also know that the unprecedented growth rate of 2% per year may be interrupted in the foreseeable future by an exhaustion of the available food supply and that the world population will then enter a lag period just prior to plunging headlong into a rapidly accelerating decline in number that can end only in its extinction.

Providing food to feed the people of the world (a population of 4.5 billion is the projected figure for the year 1976) presents a formidable and imminent problem. Man must make use of food from other than conventional sources, set up food chains that will result in high quality protein from basic food sources, and must seek to discover the mysteries of photosynthesis so that he can synthesize food substances from inorganic materials. As yet, he has discovered no workable solution to the problem.

MATERIALS REQUIRED

A project which explores the use of algae as a food source requires:

a culture of *Chlorella pyrenoidosa;*
sterile Knop's culture medium;
15 cm. culture tubes with metal closures;
sterile 1 ml. pipettes;
fluorescent and incandescent light sources;
sterile 0.5% solution of sodium bicarbonate;
a centrifuge;
centrifuge tubes; and
small laboratory animals.

DEVELOPMENT

Students can initiate the study with Chlorella, cultured in the laboratory. They may use a variation of Knop's culture medium in which the following are dissolved, separately, in 250 ml. volumes of distilled water and combined to yield one liter of mineral solution:

1 gm. KNO_3

1 gm. $MgSO_4$

1 gm. K_2HPO_4

3 gm. $Ca(NO_3)_2$

One gm. of dextrose should then be added to the mixture and the resulting media dispensed in 10 ml. volumes in metal-topped culture tubes, sterilized by autoclaving at 121° C under 15 pounds pressure for 15 minutes, and refrigerated until needed.

When beginning the project, tubes of media at room temperature should be inoculated, via sterile technique, with *Chlorella pyrenoidosa* and 4-5 drops of sterile sodium bicarbonate solution added to each. Tubes should then be placed in open racks and exposed to constant light provided by both incandescent and fluorescent sources in a protected area with a constant temperature of 72-74° F.

Cultures should be examined daily for growth and weekly sub-cultures should be made by transferring, by sterile pipette, 0.1 ml. of a thriving culture to a fresh tube of sterile media with sodium bicarbonate additive. Each week the algae should be harvested: the contents of all culture tubes should be pooled and centrifuged; the supernatant discarded; the packed cells retained for pulverization or other processing; and the total yield employed as a food supplement in the diet of small laboratory animals.

Charts should be kept to record the weekly harvests in dry weight per given number of 10 ml. tube cultures and, if a second algal species, such as *Chlorella vulgaris,* is included in the project, a comparison can be made to determine the more productive.

DISCUSSION AND INTERPRETATION

The project should be evaluated on the basis of the yield of the packed Chlorella cells and their value as food supplements for the laboratory animals. This should be determined by the weight gain, growth, state of health and vigor, and general appearance of the experimental group as compared with a control group which has received none of the supplement.

The value of using both incandescent and fluorescent light sources in the project should also be discussed, with attention given to the effects of the different wave lengths individually and in combination on normal plant growth.

Nutritional values of the Chlorella should, of course, be the major consideration. Its high protein content which has made it a popular algal form for this type of research should be noted and large batch culture methods, particularly where provisions are also made for utilization of oxygen produced, should be researched and reported for group discussion. In this relation, the implementation of these methods in space travel should be assessed.

Fish flour prepared from waste hake fish should be discussed from the standpoint of its nutritional value and the economies involved. The transformation of wastes from industry, particularly petroleum processing, should be researched and methods of obtaining usable food from these sources as well as costs, yields, and nutritional values determined.

FURTHER STUDY

The basic project is flexible and allows for many variations in materials and design. Students should be encouraged to test other algae, other photoperiods, and other sources of light, including natural daylight. They should experiment with ways of making seaweeds and algae palatable, whether to be used in unadulterated form, as supplements in food for humans, or to be used to feed animals which will, in turn, convert the basic protein to a high quality protein food for man. The highly motivated and mechanically inclined should extend the project to an investigation of methods for culturing Chlorella in a closed environmental system which might reasonably find practical use in space travel.

Advanced students should obtain strains of yeast that can secure carbon and hydrogen from wax found in petroleum and, employing techniques developed for processing petroleum with yeast organisms, rid the petroleum of the unwanted waxes while simultaneously cultivating the yeasts. The resulting protein should then be harvested and fed to small laboratory animals and the merits of the process compared with that of the basic project as a method of securing basic food.

The possibilities of some modification of the carbon cycle by chemical alterations at certain points in the process of photosynthesis will provide a challenging study for a very bright student. After researching the background chemistry, the student should plan a project in which chemicals, sprayed on suitably selected plants, induce the production of usable protein by plant structures which characteristically produce a fibrous growth, normally not usable for food.

Project—THE SEARCH FOR EFFECTIVE NON-CHEMICAL INSECTICIDES

MOTIVATION

Man's most potent weapon in his war against insects has backfired. Chlorinated hydrocarbon insecticides have polluted all parts of the earth and their persistence has been revealed in the form of harmful residues which have built up in animal and human tissues; studies of pharmacologic effects of DDT show that even minute quantities interfere with calcium metabolism necessary for bird egg shell formation and with the use of vitamin A in mammals; and conservationists continue to warn of the possible extinction of many wildlife forms if the employment of chemical poisons is not curtailed. As of now, the sale of DDT has been banned in Michigan and its use suspended in many other areas.

Nor can we maintain our disease prevention or agricultural production levels without effective pest control measures. Recently the President's Science Advisory Committee predicted that the use of insecticides must be increased from 120,000 to 700,000 metric tons annually if food production necessary to sustain our rapidly growing population is to be doubled by 1985. The question remains, "Must these insecticides be chemicals which are poisonous to life?" Researchers are hopeful that some of the non-chemical insect controls which they have been investigating will prove to be practical and effective, and without harm to other forms of life, including man.

MATERIALS REQUIRED

A project in which the employment of light as a possible agent for insect control is investigated can be performed with these simple and readily available materials:

> beetles or other insects;
> a variety of light sources: infrared, ultraviolet, and natural or white light; and
> a series of 3-compartmented, light controlled insect boxes.

DEVELOPMENT

Beetles in various stages of development can be used effectively for the determination of practical methods of insect control. They should be raised to produce, simultaneously, some larval, pupal and adult forms for study. A sampling of each of the three developmental stages should then be transferred to separate compartments in insect boxes equipped with screening to secure the specimens and exposed to the light of an infrared lamp for varying time exposures, ranging from 2-10 minutes. Similarly, insect testing boxes for larvae, pupae, and adults should be prepared for exposures to ultraviolet light and, in a control situation, to white, natural light for the same time periods.

Records of the effects of differential light wave lengths for the various time exposures should be recorded on a data chart and compared for their effectiveness as insecticides.

DISCUSSION AND INTERPRETATION

The effectiveness of light as a non-chemical insect control should be evaluated on the basis of the project findings; the effects of different wave lengths on the different stages of insect development should be determined and the stage which exhibits the greatest vulnerability identified. The practicability of using light as an agent for insect control should be assessed and the possibility of employing these wave lengths on the plants upon which the insects feed considered as an alternative method of controlling the pests.

Other non-chemical controls should also be researched and reported: the specificity of viruses in their attack on host organisms should be considered from the standpoint of locating viruses for infecting insects while remaining harmless to other animals and to man; and practices of releasing sterile male insects to mate with females and compete with fertile males should be assessed in the light of damage to vegetation during years when the practice was effective as compared with years when the practice was not employed.

The immediacy of the need to find a substitute for chemical controls should be discussed in depth. The recent report from the State University of New York at Stony Brook in which Dr. Charles F. Wurster Jr. reports findings that 0.1 ppm of DDT in sea water is sufficient to reduce the activity of phytoplankton to 10% of normal should be considered as it exerts an effect on the production of food and oxygen supplies for all earth creatures, including man.

FURTHER STUDY

Students should plan projects to extend the basic study, employing non-chemical agents such as ultrasonic waves, magnetic fields, and temperature variations as the test factors.

Those who are interested in exploring the effectiveness of chemicals in reducing the level of insecticide contamination in meat, fish, and dairy products should design a project in which chemical additives to the diet of insecticide-contaminated laboratory animals are tested. They should feed chickens or small mammals some dieldrin-sprayed feed and, in a controlled situation, a follow-up of activated charcoal to test its effectiveness in lowering the level of insecticide residues. The project might be varied to include dieldrin-contaminated aquarium water with charcoal additives in the insecticide-polluted water, or it might be used as a point of departure for embarking on a full scale project investigation into a related area of major concern—that of Environmental Pollution—its abatement and its control.

Students who are highly motivated by the published findings of studies in which hospital patients taking phenobarbitol and diphenylhydantoin averaged DDT residues between 1.7 and 1.9 parts per billion in whole blood, as compared with the 8-16 parts per billion reported for the general population in the same area, should be encouraged to pursue a research study which searches out antidotes for DDT which might be suitable for use by humans.

Project—THE SEARCH FOR A FAVORABLE EQ (ENVIRONMENTAL QUALITY)

MOTIVATION

We are becoming increasingly aware of the role man plays in the rapid deterioration of the natural environment and of the deleterious effects this will have on our very existence if immediate and drastic action is not taken. While we are properly outraged over conditions of smog, acrid odors of waste gases, accumulations of smoke and haze over the countryside, fish kills and disappearance of commercial and sport fishing areas, eradication of some of our most favored songbirds, noxious automobile exhaust fumes, pollution of many of our favorite recreational spots, and increasingly scant vegetation along stream beds near water treatment plants, we have only just begun to assess the possible long-range effects on human life due to pesticides, detergents, radiation, smoke, and industrial effluent.

In air and water pollution, industry is a chief offender and urban areas the most seriously affected. However, our zeal for cleanliness the country over is responsible for the pouring of over 3,000,000,000 pounds of synthetic detergents annually down the nation's household drains. Unquestionably they are effective as cleansing agents, but the synthetic ingredients of these detergent preparations are not readily degraded by conventional waste water treatment procedures; hence they contribute significantly to the problem of pollution. Their return to households in the form of "foaming" tap water and their effects on food chains, with man at the top of the chain, have serious implications for us all.

It is imperative that we encourage further development of the technologies that will correct problems dealing with the pollution of our environment and active involvement in programs which seek to bring pressures to bear on those who would show wilful and flagrant disregard for its quality. It is a job for all, NOW.

MATERIALS REQUIRED

A project which investigates the effects of detergent solutions on plant life can be conducted in a simple manner and will require the following:

500 viable lima bean seeds;

a 2% solution of sodium hypochlorite;

household laundry or dishwashing detergent solutions:
0.1%, 0.5%, 1.0%, and 5.0%;

a good garden or potting soil mixture;

detergent-free water;

five seed planters of suitable size to accommodate 100 seeds each; and

a growing area which provides suitable conditions of air, light, and temperature.

DEVELOPMENT

Designed to help students develop an understanding of the effects of pollution on life forms and an awareness of the need for a quality environment, the project should be conducted with young seedlings grown under experimental conditions of varying concentrations of detergent in the water used to supply them. Five hundred seeds that have been previously soaked for ten minutes in a 2% solution of sodium hypochlorite and one hour in detergent-free water should be planted in sets of one hundred in identical growth chambers and designated as sets A, B, C, D, and E. Keeping all other factors constant, set A should be watered daily with a 0.1% detergent solution, set B with a 0.5% solution, set C with a 1.0% solution, set D with a 5% solution, and set E, acting as a control, with detergent-free water.

Observations concerning the number of seeds germinated, the average height of seedlings, and the general appearance and number of living seedlings in each group should be made daily over a 10-12 day period and recorded on a suitably constructed data chart. Using a previously determined color code to identify the differentially treated sets of seedlings, the number of living plants in each set should be plotted on the same axis for a comparative study of plant growth accompanying the use of detergent-contaminated and detergent-free water.

DISCUSSION AND INTERPRETATION

Questions should be posed for students to ponder: "What effect do detergents exert on plant growth?" "At what concentrations can detergents be tolerated by plants?" "What is the mode of action of detergent substances as they inhibit the growth of plants?" "How does this affect animals and man?"

The data collected should be analyzed and the relationship between success of plant life and the degree of detergent pollution in the water determined by an interpretation of the growth rate curves plotted and the physical appearance of the seedlings in each set over the period of time they were being treated. The number of survivors remaining at the end of the 10-12 day period should be viewed as an indication of the inhibitory action of the various concentration levels of the detergent in water.

To determine the mode of action of the detergent in inhibiting plant growth, a highly motivated student can perform a demonstration to the class using a simple procedure. Place five identical sprigs of Elodea in separate beakers containing, respectively, detergent-free water and 0.1%, 0.5%, 1.0% and 5.0% detergent solutions; after having placed them for several hours in bright sunlight with narrow graduated cylinders positioned directly over the plants, collect and measure the amount of oyxgen produced by photosynthesis in each media; examine leaf cells from each sprig under the high power lens of the microscope and compare the number and appearance of chloroplasts in leaf cells from plants engaging vigorously in the photosynthetic process with those from plants in which the process was inhibited; identify plasmolyzed cells and relate to the detergent solution concentration in which they were produced and to the inhibition of lima bean seedling growth as observed in the project.

How the effects of water pollution on other forms of life in a food web are brought about should also be discussed, with cases where a specific pollutant, such as DDT, are cited. The complex pathways along which food is passed should be traced from plant to plant-eating animal to carnivore, with an accompanying increase in concentration of the pollutant in body tissues as it moves through the web. Analyses indicating that predaceous fish contain a higher level of DDT than do those that feed on plankton should be researched, and how DDT at low concentrations found to be effective when used as an insecticide builds up to 25% and greater in some fish-eating birds should be determined. In this relation, the manner in which a build-up of 12 ppm. of DDT found to be present in the average human should be discussed and reconciled with the 5 ppm. set as the maximum amount allowable in fish sold commercially. Students might also consider the prognosis that residual DDT collecting in the human liver will soon make dangerous the use of certain common drugs and increase the incidence of cancer of the liver.

Other phases of pollution—noise, industrial effluent, cigarettes, automobile exhaust, commercial and industrial blight, and radio-active waste in the aquatic environment—should also be considered, with attention given to the social and political overtones which are associated with the biological problems. The events of the Torrey Canyon and Santa Barbara disasters, the Minamata Bay tragedy, the Escombia Bay fish kill, the "dead sea" created in the dumping grounds off Ambrose Light, the death of Lake Erie, and the increased death rates during periods of heavy smog experienced in London, the Meuse Valley in Belgium and Donora, Pennsylvania, all should receive due attention, with the economic losses as well as the health and esthetic facets of environmental pollution being considered. On the positive side, currently available and projected methods of pollution control should be researched and the progress being made in combating pollution problems, such as in Pittsburgh, Pennsylvania, should be recognized. The historical as well as the current aspects of our still growing pollution problems should be assessed and the present day application of the method expressed in the old adage, "The solution to pollution is dilution," debated in class.

Predictions for the near future—that air pollution will have reduced by one-half the amount of sunlight reaching the earth, that urban dwellers will be forced to wear gas masks in order to survive, that increased CO_2 in the atmosphere will affect the earth's temperature and lead to mass flooding or a new ice age, and that the breakdown of a major ecological system will bring about new diseases to which humans have no resistance—should be examined objectively and critically.

Nor should the immediacy of pollution problems in the community be ignored. Where applicable, active involvement in a community study of stream pollution, a local clean-up campaign, or a community survey patterned after the recent Gallup national public opinion poll which revealed that 73% of all Americans are willing to pay additional taxes if earmarked specifically for the improvement of their natural surroundings, sponsorship of a poster and/or photo contest to point up local problem areas, and the organization of an "Environmental Teach-In," should be encouraged. Thus aroused, students can develop both an insight into the enormity and urgency of the problems associated with our rapidly deteriorating natural environment and a positive attitude toward implementing effective, rigorous, and on-going corrective programs NOW.

FURTHER STUDY

The project can be repeated, employing a wider or narrower

range of detergent concentrations, a variety of household detergent preparations, or seedlings of a different plant species as the experimental organisms. The design of the project also can be varied; some students should be encouraged to plan a project study of the relationship between detergent concentrations and the metabolism of organisms, using yeast cells in suspension as the experimental organisms and CO_2 production as an indication of their rates of metabolism.

Those students wishing to extend the basic project might well investigate the tolerance level of a species of aquatic plant or animal life and the transfer of harmful effects, through build-up, to animals feeding on these as basic food. Additionally, they should research species of bacteria reputed to be capable of degrading detergent molecules and, in a project employing such microorganisms, attempt to restore detergent-polluted water to a condition of safety for use by plant, animal, and human life forms. In another study, students should design a project which investigates the poisonous effects on man due to nitrate-polluted ground water from chemical fertilizers. In still another, they should search out possible methods of reducing the insecticide uptake in plant crops, employing pea plants grown in insecticide-contaminated soil, with and without charcoal additives. For a challenging study involving animals, students should be encouraged to plan and conduct a project in which they explore the possible damage to shells of birds' eggs caused by PCB, a waste formed when plastics are burned. Using doves as the experimental organisms, they should raise both a control and an experimental group for a comparative study and assessment of the long-range effects of such substances on bird populations.

A project which seeks to determine the effects of ozone on plant life should be recommended to those who exhibit a talent for ingenious project design. Employing UV light (for the irradiation of air to form ozone) in contact with petunia plants grown in an airtight glass container, they should assess the oxidant damage to the leaves of the experimental plants and, for a special challenge, endeavor to locate or develop a strain of the same plant which is immune to the ozone effects.

Students should also be encouraged to investigate methods for detection and analysis of air and water pollutants and to demonstrate these techniques to the class. Also, goldfish should be employed to detect water pollution by their behavior patterns and the resulting bioassay compared with the chemical methods used for an assessment of its reliability and sensitivity. Those wishing to investigate the effects of tobacco smoke on small animals should devise a smoking mechanism by which the animals will be forced to inhale tobacco smoke and, using littermate white rats, run a comparative study where some are exposed to the smoke and others, unexposed, are used as a control. With modifications in design, the mechanism can be used in a project study investigating the effects of oxides of nitrogen and sulfur and some of the hydrocarbon gases which commonly pollute the atmosphere.

For an entire class field project, an ecological study of a stream or river in the community should be undertaken. Beginning from a point upstream, the river should be followed for a mile or more downstream, during which conditions of the water and of the plant and animal life should be entered on a map. Signs of pollution should be noted and the sources, if possible, traced. Such a map, accompanied by photographs, would make an excellent bulletin board or showcase display for creating a greater awareness of the enormity of the problems of pollution and of their relevancy to the community.

Project—THE SEARCH FOR WAYS TO INCREASE MAN'S LONGEVITY

MOTIVATION

A perusal of both historical accounts and some present day health regimes and commercially sponsored products and treatments will attest to man's eternal quest for a "Fountain of Youth." Graying of hair, loss of elasticity of skin and bone, replacement of muscle by connective tissue, slowing of reaction time, and a degree of senility are symptoms of the aging process which, due to phenomenal medical and technological progress resulting in more humans attaining a full life expectancy of about 80 years, are being experienced by a constantly increasing percentage of the population. But the aging process occurs independently of accident and disease, and man's life expectancy today is still not appreciably greater than it was foretold in the Bible at "three score and ten."

Presently, gerontologists are exploring genetic as well as biochemical parameters of aging and, based on their experimentation with animals, have proposed several theories as to how the aging process occurs and what scientific regimes hold promise for increasing man's longevity. The mystery of aging is truly one of biology's most baffling and challenging problems.

MATERIALS REQUIRED

The effects of diet on longevity can be studied with the following few materials and simple laboratory equipment:
   littermate laboratory rats;
   differential rat diets (a normal laboratory rat diet preparation and a starvation rat diet preparation);
   rat cages;
   animal water bottles; and
   animal weighing scales

DEVELOPMENT

A project which undertakes the determination of longevity in laboratory rats is, of necessity, a long-range study starting with the separation of weanling littermate rats into two groups, each to receive the finest of laboratory care and handling which must be identical except for the factor of diet. The control group should be fed a normal laboratory rat diet and the experimental group essentially the same but without carbohydrates and fats. The experimental diet should be limited to proteins, minerals, and vitamins.

Animals should be cared for and fed daily, with weekly tabulations of average weight, general appearance, and activity for each group recorded for the duration of the project. Long-term results, as determined by evidence of maturation and aging, should be determined periodically and at the end of a 24-26 month period.

DISCUSSION AND INTERPRETATION

Students who initiate the project should represent several grade levels, and records, as well as the project, should be continuous and on-going to facilitate a proper interpretation.

The effects of diet on aging should be the primary consideration of the study. Students who view the effects of early maturation and aging should relate it to diet and associate this observation with our current practices of infant and child feeding. They should discuss the recent lowering of daily calorie recommendations made by the Food and Nutrition Board of the National Research Council and, for a spirited class debate, be offered the topic, "The Psychology of Modern Infant Feeding Properly Belongs to the Butcher."

The biochemistry of aging should be researched and discussed as well as theories which link somatic mutations to the aging process and relationships that have been observed between radiation and aging. Students should also give consideration to the theory that changes accompanying the aging process occur in cell nuclei and relate this theory to difficulties they may have encountered in initiating cell cultures from adult tissues as contrasted with the relative ease with which cells from embryonic tissues are cultured *in vitro*.

The relative rates of aging observed to occur in men and women should receive attention, with students being encouraged to offer plausible reasons for the differential in physiological degeneration. Considering man's eradication of infectious diseases and inhibition of the progression of degenerative diseases, students should ponder the question if man should not plan early for a healthy old age to rival the 120 year life expectancy enjoyed by the fabled Hunzukuts of northern Pakistan.

FURTHER STUDY

Students should explore additional scientific regimes for a determination of other factors which may affect the aging process. Some might design projects in which the effects of chemicals, such as magnesium chloride, on the aging of guppies and/or goldfish are determined while others pursue an investigation of the effects of anti-radiation drugs, such as cysteine, on the longevity of mice in the laboratory. The juvenile hormone in insects should also be researched and a project study, using both larval and pupal stages of insects, be conducted to determine relationships between injections of this substance at various stages of organismal development and the total aging process.

Students with a strong background in chemistry should be challenged to search out and test free radicals which might be expected to combine chemically with the highly reactive OH⁻ molecular fragments characteristically found in older cells. Using this approach, with cells cultured *in vitro* or with experimental mice, they should speculate on the practicability of using chemicals to inhibit activity of molecular fragments that would normally play a part in cell deterioration and of their effectiveness in delaying the process of aging in animals and man.

Students who are motivated by the observation that cold-blooded animals, such as tortoises, usually enjoy a greater longevity than do animals which are warm-blooded should be encouraged to design and conduct a project in which they investigate the effects of lowering the body temperature of warm-blooded animals in the laboratory. Using hamsters or white mice as the experimental organisms, they should test the effects of drugs that reduce the body temperature within a range of 2-3 degrees.

In whatever project activity undertaken in this general topic of study, students should be advised of proper techniques for care and handling of experimental animals used and should be encouraged to employ procedures which are scientifically sound while pursuing investigations based on original ideas and creative designs.

RECOMMENDED READING

Boehm, George A.W. "The Search for Ways to Keep Youthful," *Fortune*, LXXI, (March, 1965), 3.

Brown, Harrison. *The Challenge of Man's Future*. Viking Press, New York, 1956.

Brown, Robert E. "Detection and Analysis of Air and Water Pollutants," *The American Biology Teacher*, (1969), 31, 5.

Butler, Philip A. "Monitoring Pesticide Pollution," *BioScience*, (1969), 19, 10.

Carson, Rachel. *Silent Spring.* Houghton Mifflin Company, Boston, 1962

Cole, L.C. "Thermal Pollution," *BioScience,* (1969), 19, 11.

Comfort, Alex. "The Life Span of Animals," *Scientific American,* (1961), 205, 2.

Culliton, Barbara J. "Cell Cycles Give Clues," *Science News,* (1968), 94, 21.

———. "Safety Challenged," *Science News,* (1968), 94, 26.

Curtis, Howard J. "Biological Mechanisms Underlying the Aging Process," *Science,* (August, 1963), 141.

Cusack, Michael. "Nitrates—Poisons From a Faucet," *Science World,* (1970), 20, 4.

Devlin, John C. "Prying Medical Secrets from Creatures of the Sea," *Welch Biology and General Science Digest,* (1966), 16, 1.

DuBridge, Lee A. "Science Serves Society," *Science,* (1969), 164, 3884.

Faltermayer, Edmund K. "We Can Afford Clean Air," *Fortune,* LXXII, (1965), 5.

Fisher, Arthur. "Radiation and Life," *Senior Science,* (February, 1965).

Friedman, Sharon. "Man's Survival in a Changing World," *BioScience,* (1968), 18, 10.

Gillon, Hadassah. "Tissue Survival Prolonged," *Science News,* (1968), 94, 5.

Georgopoulos, S.G. "The Problem of Fungicide Resistance," *BioScience,* (1969), 19, 11.

Guinness, Alma. "Wastes and Water: Perilous Balance," *Science World,* (1965), 17, 8.

Hamilton, Andrew. "How We Can Delay Old Age," *Science Digest,* (1966), 59, 2.

Hayflick, Leonard. "Human Cells and Aging," *Scientific American,* (1968), 218, 3.

Hennigan, R.D. "Water Pollution," *BioScience,* (1969), 19, 11.

Howard, Walter E. "The Population Crisis Is Here Now," *BioScience,* (1969), 19, 9.

Hulett, H.R. "Optimum World Population," *BioScience,* (1970), 20, 3.

Johnson, S.P. and J.C. Finn, Jr. "Ecological Considerations of a Permanent Lunar Base," *The American Biology Teacher,* (1963), 25, 7.

Kimball, Thomas L. "Our National EQ," *National Wildlife,* (1969), 7, 5.

Leskowitz, Sidney. "Immunologic Tolerance," *BioScience,* (1968), 18, 11.

Lorant, Bernard. *Will the Meek Inherit the Earth?* Velsicol Chemical Corporation, Chicago, Illinois, 1966.

Matthern, Robert O. and Robert B. Kich. "The Continuous Culture of Algae Under High Light Intensity," *The American Biology Teacher,* (1963), 27, 7.

McElroy, William D. "Biomedical Aspects of Population Control," *BioScience,* (1969), 19, 1.

Menaker, M. "Biological Clocks," *BioScience,* (1969), 19, 8.

New, John G. and J. Gary Holway. 'Pollution—Is There a Solution?" *BioScience,* (1969) 19, 10.

Peakall, David B. "Pesticides and the Reproduction of Birds", *Scientific American,* (1970), 222, 4.

Pirie, N.W. "Orthodox and Unorthodox Methods of Meeting World Food Needs," *Scientific American,* (1967), 216, 2.

Proujan, Carl. "Fresh Water for the Future: Flood or Trickle?" *Science World,* (1965), 18, 8.

Puck, Theodore T. "Radiation and the Human Cell," *Scientific American,* (1960), 202, 4.

Rutzler, Klaus and Wolfgang Sterrer. "Oil Pollution," *BioScience,* (1970), 20, 4.

Sax, Karl. *Standing Room Only.* Beacon Press, Boston, 1955.

Scheinfeld, Amram. "The Mortality of Men and Women," *Scientific American,* (1958), 198, 2.

Shock, Nathan W. "The Physiology of Aging," *Scientific American,* (1962), 206, 1.

Shumway, N., F. Marley, C. Weathersbee, C. Behrens, P. McBroom, and B. Culliton. "Transplants—The State of the Art," *Science News,* (1968), 93, 9.

Silva, James. "Food, Space, and Survival," *Science World,* (1963), 13, 8.

————. "Food, Men, and Microbes," *Senior Science,* (1965), 17, 5.

Sinex, F.M. "Biochemistry of Aging," *Science,* (1961), 134, 3488.

Swisher, R.D. "Detergent Enzymes: Biodegradation and Environmental Acceptability," *BioScience,* (1969), 19, 12.

Verzar, Frederic. "The Aging of Collagen," *Scientific American,* (1963), 208, 4.

Wallace, B. and T. Dobzhansky, *Radiation , Genes, and Man.* Rinehart and Winston, New York, 1963.

Ward, C.H., S.S. Wilks, and H.L. Croft. "The Use of Algae and Other Plants in the Development of Life Support Systems," *The American Biology Teacher,* (1963), 25, 7.

Weathersbee, Christopher. "Convert Newsprint to Steak," *Science News,* (1968), 94, 9.

———— "Antigens vs. Antibodies, *Science News,* (1968), 93, 27.

Weiss, Malcolm. "DDT—Misguided Missile?" *Science World,* (1970), 20, 3.

Williamson, F.S.L. "Population Pollution, " *BioScience,* (1969), 19, 11.

Wite-Stevens, Robert. *The Facts and Fallacies of Silent Spring*, American Cyanimid Co., Princeton, N. J., 1962.

Wroblewski, Felix. "Enzymes in Medical Diagnosis," *Scientific American*, (1961), 205, 2.

Wurster, Charles F. "DDT Goes to Trial in Madison," *BioScience*, (1969). 19, 9.

Index